U0327593

本书收录部分中西部地区县级综合档案馆一览

18 省 5 自治区 1 直辖市 1 兵团

70 家场馆 / 占地总面积 46.6 万 m^2

建筑总面积 39.6 万 m^2

库房总面积 15.3 万 m^2

代表档案馆数量

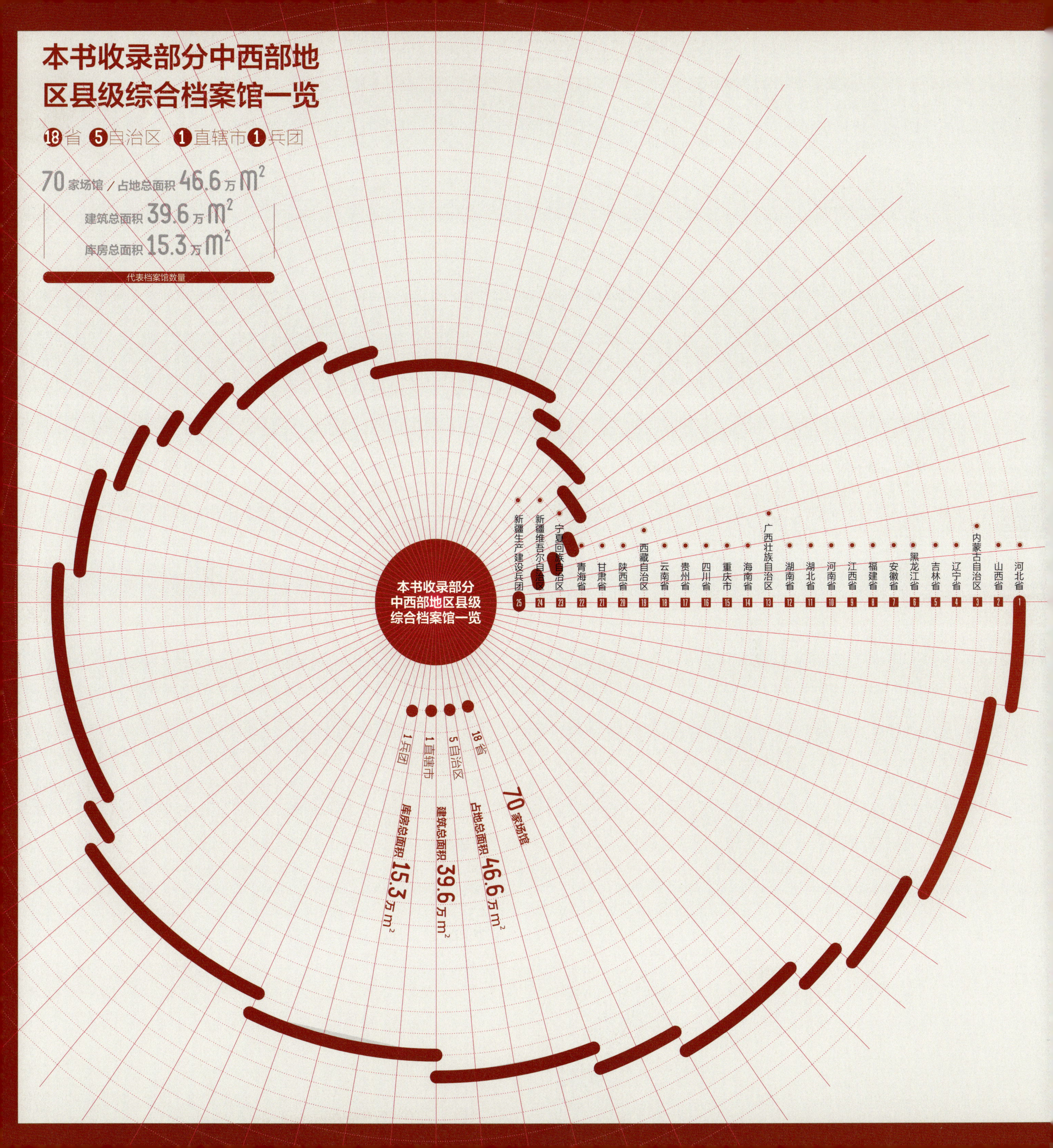

01 河北省

唐山市丰南区档案馆
占地面积：5600m²
建筑面积：6294m²
库房面积：2200m²
层　　数：地上 4 层，地下 1 层
建成时间：2011 年 3 月

献县档案馆
占地面积：10000m²
建筑面积：4439m²
库房面积：1484m²
层　　数：4 层
建成时间：2015 年 12 月

02 山西省

高平市档案馆
占地面积：1242m²
建筑面积：6438m²
库房面积：2010m²
层　　数：地上 5 层，地下 1 层
建成时间：2016 年 5 月

沁县档案馆
占地面积：3266m²
建筑面积：3052m²
库房面积：1067m²
层　　数：主楼 3 层
建成时间：2017 年 11 月

阳泉市郊区档案馆
占地面积：3346m²
建筑面积：5127m²
库房面积：2082m²
层　　数：主楼 6 层
建成时间：2012 年 12 月

垣曲县档案馆
占地面积：3400m²
建筑面积：3560m²
库房面积：1260m²
层　　数：地上 5 层，地下 1 层
建成时间：2016 年 5 月

03 内蒙古自治区

鄂尔多斯市东胜区档案馆
占地面积：55915m²
建筑面积：5973m²
库房面积：2620m²
层　　数：地上 2 层，地下 1 层
建成时间：2014 年 11 月

新巴尔虎右旗档案馆
占地面积：4200m²
建筑面积：3007m²
库房面积：980m²
层　　数：4 层
建成时间：2015 年 1 月

04 辽宁省

沈阳市于洪区档案馆
占地面积：4235m²
建筑面积：5221m²
库房面积：1500m²
层　　数：主体 5 层
建成时间：2012 年 3 月

05 吉林省

珲春市档案馆
占地面积：6000m²
建筑面积：5786m²
库房面积：1984m²
层　　数：6 层
建成时间：2016 年 11 月

龙井市档案馆
占地面积：9098m²
建筑面积：4680m²
库房面积：1733m²
层　　数：主楼 4 层
建成时间：2018 年 12 月

通化县档案馆
占地面积：1569m²
建筑面积：5087m²
库房面积：2687m²
层　　数：6 层
建成时间：2018 年 11 月

06 黑龙江省

甘南县档案馆
占地面积：2500m²
建筑面积：3642m²
库房面积：2105m²
层　　数：主楼 4 层、配楼 1 层
建成时间：2016 年 10 月

桦南县档案馆
占地面积：2272m²
建筑面积：4019m²
库房面积：1434m²
层　　数：5 层
建成时间：2017 年 11 月

07 安徽省

霍山县档案馆
占地面积：4467m²
建筑面积：4699m²
库房面积：1853m²
层　　数：5 层
建成时间：2011 年 12 月

青阳县档案馆
占地面积：8800m²
建筑面积：3563m²
库房面积：1270m²
层　　数：2 层
建成时间：2018 年 8 月

泗县档案馆
占地面积：10000m²
建筑面积：7298m²
库房面积：1884m²
层　　数：主楼 4 层
建成时间：2018 年 7 月

望江县档案馆
占地面积：14674m²
建筑面积：5592m²
库房面积：2174m²
层　　数：地上 3 层，地下 1 层
建成时间：2013 年 11 月

08 福建省

福清市档案馆
占地面积：20000m²
建筑面积：11450m²
库房面积：2604m²
层　　数：地上 10 层，地下 1 层
建成时间：2016 年 8 月

晋江市档案馆
占地面积：7402m²
建筑面积：17680m²
库房面积：6000m²
层　　数：地上 9 层，地下 1 层
建成时间：2013 年 6 月

罗源县档案馆
占地面积：4203m²
建筑面积：3997m²
库房面积：1326m²
层　　数：主楼 5 层
建成时间：2017 年 12 月

上杭县档案馆
占地面积：6500m²
建筑面积：8938m²
库房面积：3000m²
层　　数：主体 3 层
建成时间：2018 年 3 月

武平县档案馆
占地面积：3333m²
建筑面积：4600m²
库房面积：2643m²
层　　数：主楼 5 层
建成时间：2016 年 1 月

09 江西省

崇义县档案馆
占地面积：8000m²
建筑面积：2839m²
库房面积：1500m²
层　　数：2 层
建成时间：2018 年 2 月

高安市档案馆
占地面积：4200m²
建筑面积：12560m²
库房面积：4000m²
层　　数：地上 4 层，地下 1 层
建成时间：2018 年 1 月

井冈山市档案馆
占地面积：933m²
建筑面积：3868m²
库房面积：1600m²
层　　数：4 层
建成时间：2016 年 11 月

万载县档案馆
占地面积：9986m²
建筑面积：4700m²
库房面积：1465m²
层　　数：5 层
建成时间：2015 年 10 月

武宁县档案馆
占地面积：972m²
建筑面积：4431m²
库房面积：1772m²
层　　数：5 层
建成时间：2014 年 10 月

修水县档案馆
占地面积：1814m²
建筑面积：5329m²
库房面积：2200m²
层　　数：3 层
建成时间：2013 年 4 月

10 河南省

睢县档案馆
占地面积：2150m²
建筑面积：5000m²
库房面积：1646m²
层　　数：3 层
建成时间：2017 年 7 月

11 湖北省

红安县档案馆
占地面积：16717m²
建筑面积：7433m²
库房面积：2709m²
层　　数：4 层
建成时间：2012 年 12 月

黄梅县档案馆
占地面积：5788m²
建筑面积：5589m²
库房面积：2166m²
层　　数：3 层
建成时间：2016 年 7 月

建始县档案馆
占地面积：1600m²
建筑面积：6304m²
库房面积：2400m²
层　　数：7 层
建成时间：2012 年 6 月

利川市档案馆
占地面积：8000m²
建筑面积：6775m²
库房面积：2200m²
层　　数：地上 6 层，地下 1 层
建成时间：2017 年 12 月

松滋市档案馆
占地面积：1056m²
建筑面积：4054m²
库房面积：1488m²
层　　数：4 层
建成时间：2011 年 6 月

咸宁市咸安区档案馆
占地面积：10000m²
建筑面积：4954m²
库房面积：1833m²
层　　数：4 层
建成时间：2017 年 12 月

阳新县档案馆
占地面积：6670m²
建筑面积：5579m²
库房面积：2095m²
层　　数：6 层
建成时间：2015 年 7 月

12 湖南省

	浏阳市档案馆	平江县档案馆	湘潭县档案馆
占地面积	7630m^2	13474m^2	5420m^2
建筑面积	7081m^2	16528m^2	13521m^2
库房面积	3015m^2	4700m^2	6800m^2
层　　数	主楼 6 层，配楼 4 层，地下 1 层	主楼 6 层、配楼 3 层、地下 1 层	主楼 10 层，地下 1 层
建成时间	2012 年 12 月	2015 年 1 月	2017 年 1 月

13 广西壮族自治区

	荔浦县档案馆	南宁市西乡塘区档案馆
占地面积	1957m^2	3333m^2
建筑面积	3665m^2	5200m^2
库房面积	1102m^2	1400m^2
层　　数	5 层	主楼 5 层，地下 1 层
建成时间	2015 年 3 月	2017 年 6 月

14 海南省

	文昌市档案馆
占地面积	6590m^2
建筑面积	3709m^2
库房面积	1002m^2
层　　数	3 层
建成时间	2015 年 1 月

15 重庆市

	涪陵区档案馆	合川区档案馆
占地面积	6660m^2	6352m^2
建筑面积	7378m^2	9632m^2
库房面积	3000m^2	4000m^2
层　　数	主楼 5 层，地下 1 层	主楼 15 层，配楼 2 层，地下 1 层
建成时间	2012 年 9 月	2016 年 9 月

16 四川省

	成都市郫都区档案馆	德昌县档案馆	富顺县档案馆	米易县档案馆
占地面积	6047m^2	4230m^2	10005m^2	3335m^2
建筑面积	3991m^2	4583m^2	6800m^2	6650m^2
库房面积	2300m^2	1746m^2	2800m^2	1917m^2
层　　数	5 层	主楼 5 层，配楼 4 层	主楼 4 层，配楼 2 层	主楼 6 层，地下 1 层
建成时间	2012 年 9 月	2017 年 3 月	2017 年 12 月	2012 年 5 月

17 贵州省

	瓮安县档案馆	遵义市汇川区档案馆
占地面积	968m^2	10000m^2
建筑面积	2850m^2	10857m^2
库房面积	964m^2	3628m^2
层　　数	5 层	7 层
建成时间	2017 年 8 月	2017 年 2 月

18 云南省

	沧源佤族自治县档案馆	昌宁县档案馆	楚雄市档案馆	澜沧拉祜族自治县档案馆	陇川县档案馆	勐腊县档案馆	巍山彝族回族自治县档案馆	西盟佤族自治县档案馆	永德县档案馆
占地面积	4725m^2	6666m^2	11774m^2	3333 m^2	745m^2	3333m^2	3086m^2	3973m^2	3833m^2
建筑面积	3675m^2	5017m^2	11068m^2	4643 m^2	3602m^2	4513m^2	4769m^2	3281m^2	5056m^2
库房面积	1785m^2	1407m^2	5574m^2	2389 m^2	1800m^2	2149m^2	2381m^2	1550m^2	2587m^2
层　　数	5 层	4 层	6 层	6 层	5 层	4 层	5 层	主楼 3 层，配楼 2 层	5 层
建成时间	2016 年 6 月	2015 年 12 月	2012 年 10 月	2016 年 10 月	2015 年 5 月	2014 年 10 月	2014 年 2 月	2011 年 11 月	2015 年 10 月

19 西藏自治区

	墨脱县档案馆
占地面积	3000m^2
建筑面积	895m^2
库房面积	545m^2
层　　数	6 层
建成时间	2014 年 5 月

20 陕西省

	长武县档案馆	合阳县档案馆	山阳县档案馆
占地面积	8026m^2	2520m^2	3333m^2
建筑面积	7068m^2	3326m^2	4359m^2
库房面积	1993m^2	1400m^2	1482m^2
层　　数	地上 4 层，地下 1 层	主楼 5 层，配楼 1 层	主楼 5 层，配楼 4 层
建成时间	2016 年 3 月	2018 年 5 月	2014 年 6 月

21 甘肃省

	敦煌市档案馆	金昌市金川区档案馆
占地面积	20244m^2	14500m^2
建筑面积	4177m^2	3740m^2
库房面积	1500m^2	2100m^2
层　　数	3 层	3 层
建成时间	2016 年 8 月	2016 年 8 月

22 青海省

	刚察县档案馆
占地面积	1500m^2
建筑面积	1398m^2
库房面积	780m^2
层　　数	主楼 3 层，配楼 1 层
建成时间	2012 年 10 月

23 宁夏回族自治区

	盐池县档案馆
占地面积	5286m^2
建筑面积	2224m^2
库房面积	916m^2
层　　数	主楼 2 层，配楼 1 层
建成时间	2018 年 12 月

24 新疆维吾尔自治区

	库尔勒市档案馆
占地面积	2560m^2
建筑面积	3745m^2
库房面积	1220m^2
层　　数	主楼 4 层，地下 1 层
建成时间	2014 年 10 月

25 新疆生产建设兵团

	北屯市档案馆
占地面积	8318m^2
建筑面积	4118m^2
库房面积	1685m^2
层　　数	地上 3 层
建成时间	2013 年 11 月

旧貌变新颜

中西部地区县级综合档案馆建筑集

Architecture Collection of the County-level General Archives in Central and Western China

国家档案局　编

Edited by National Archives Administration of China

中国建筑工业出版社

图书在版编目（CIP）数据

旧貌变新颜 中西部地区县级综合档案馆建筑集：汉英对照／国家档案局编. —北京：中国建筑工业出版社，2019.8

ISBN 978-7-112-24028-9

Ⅰ. ① 旧… Ⅱ. ① 国… Ⅲ. ① 档案馆–建筑设计–中国–图集 Ⅳ. ① TU242.3-64

中国版本图书馆CIP数据核字（2019）第157650号

责任编辑：刘　丹　徐　冉
书籍设计：锋尚设计
责任校对：王　烨

旧貌变新颜　中西部地区县级综合档案馆建筑集
国家档案局　编
*
中国建筑工业出版社出版、发行（北京海淀三里河路9号）
各地新华书店、建筑书店经销
北京锋尚制版有限公司制版
北京雅昌艺术印刷有限公司印刷
*
开本：889×1194毫米　1/12　印张：27　字数：630千字
2019年12月第一版　　2019年12月第一次印刷
定价：368.00元
ISBN 978-7-112-24028-9
（34534）

前言

档案，作为人类社会活动的真实记录，是社会进步和文明发展的见证，是把人类社会的过去、现在和将来紧紧联系在一起的纽带。没有了档案，人类的历史就会产生空白，人类的记忆就可能被抹去。作为档案的收集者、保管者和开发者，档案工作者既承担着守护历史文化遗产的责任，也承担着构建未来社会记忆的责任。这就要求档案工作者必须要有对历史负责、为现实服务、替未来着想的使命感。也正是这种责任和使命，决定了档案事业是一项崇高的事业、永恒的事业、前景光明的事业。

中国是一个有着悠久历史和灿烂文明的国家，从3000多年前的殷商甲骨文开始，我们的先人就已经开始形成并保存档案，这种重视档案的优良传统绵延几千年，一直延续到今天，成为中华民族文化得以世代传承的重要原因之一。中华人民共和国成立以来，中国政府始终把档案视为国家各项工作和人民群众各方面情况的真实记录，视为促进国家各项事业科学发展、维护国家及人民群众根本利益的重要依据；把档案工作当作国家各项工作中不可缺少的基础性工作，当作维护国家历史真实面貌的重要事业。多年来，中国各级政府通过把档案事业发展列入政府的议事日程、纳入国民经济和社会发展规划，从而保证了档案事业的持续发展。

各级各类档案馆作为永久保管各类重要档案资料的基地以及科学研究和各方面利用档案史料的中心，一直以来都是中国档案事业的主体。改革开放40年来，随着中国经济社会的快速发展，中国各级政府进一步加大了对档案事业特别是档案馆工作的支持力度。尤其是从2010年开始，国家发展改革委、国家档案局开始实施《中西部地区县级综合档案馆建设规划》（后文简称为《规划》），以中央财政投资补助的形式帮助中西部地区馆舍面积不足的县级综合档案馆对馆库进行新建或改造，以改善中西部地区基层综合档案馆建设与国内其他地区相比较为滞后的状况。截至2019年，中央投资补助资金69亿元，共实施县级档案馆建设项目1500多个。随着列入《规划》的档案馆新馆的建成和投入使用，中央投资的良好效益已显著呈现，长期以来存在于中西部地区县级综合档案馆的“无馆舍”“危房馆”现象得以彻底解决，新的档案馆库房还普遍安装了安全和消防设施，进一步消除了档案安全隐患。档案馆条件的根本改善，使社会各方面形成的档案能够被依法接收并得到妥善保管。

今天，这些新建成的档案馆由于承担了更多的社会功能，因此与过去的旧馆相比，有着更为合理的内部空间布局，包括更加友好的展览大厅、更加舒适的档案查阅区域等，一些档案馆还通过更加人性化的设计，包括对中庭空间的内涵开发、馆内外休闲空间的营造等，形成了更加亲和自然的氛围。依托良好的场馆条件，档案馆如今能够为政府和人民群众提供更加丰富的档案服务，包括提供更便捷的档案利用服务、举办更精彩的档案文化展览等，从而成为各级政府联系人民群众的桥梁和落实各项惠民政策的助手。此外，这些新建成的档案馆在建筑外观上与旧馆相比也更具特色，有的充分体现了当地的地域文化特色，与周边既有建筑和谐相处，有的在新开辟的空间环境中精心塑造档案特色，从而以自身的设计提升了城市的品位。从某种程度上说，正是这些长久矗立在城市中心的档案馆大大提高了全社会对档案工作的认知。

值此新中国成立70周年之际，国家档案局从近年来在中央投资补助下新建成的中西部地区县级综合档案馆中遴选出70家优秀实例和70家代表实例结集成书，礼赞70年来中国档案事业发展的辉煌成就，致敬新中国70周年华诞。回望历史，档案工作从来没有像今天这样如此深入地渗透到经济社会发展的各领域和各环节，从来没有像今天这样如此深刻地影响到广大人民群众的物质生活和精神生活。我们相信，未来档案和档案工作将愈来愈显示它的重要地位和作用，档案馆工作也将在“走向依法管理、走向开放、走向现代化”的道路上标注新的高度。

Foreword

Archives, being faithful recorder of social activities, are witnesses to social progress and civilization development. They link the past, present and future of human society inextricably. Without archives, human history may go blank and human memory may be erased. As collectors, custodians and developers of archives, archivists bear the responsibility of both guarding historical and cultural heritages and making memories for the future society. In this regard, archivists must be aware of their mission to stay true to history, serve the reality, and concern about the future. With this responsibility and mission in mind, archiving itself stands as a lofty, everlasting and promising cause.

China boasts of a long-standing history and a splendid civilization. The earliest form of archives in China can be traced back to the Shang Period over 3,000 years ago when our ancestors made inscriptions on bones or tortoise shells for keeping archives. The tradition of valuing archives has lasted for thousands of years, till now and is a main contributor to pass down Chinese culture generation to generation. Since the founding of the People's Republic of China, the Chinese government has been taking archives as true records about various aspects of the country and its people, and an important basis for promoting the scientific development of all undertakings of the country and safeguarding the fundamental interests of the country and its people. Archiving is also regarded as an indispensable groundwork to carry out other national tasks, as well as a key undertaking to maintain the accuracy of the country's history. Over the years, the archives undertaking in China has gone through continued development as governments at all levels have included it into the government agenda and the national economic and social development plan.

As bases for permanent custody of all kinds of important archive files and centers for scientific researches and historical documents, archival facilities of all classes and at all levels have always been the principal part in the archiving cause of China. Accompanying the rapid economic and social expansion in China since China's economic reform 40 years ago, governments at all levels have given more support to the archives cause, especially archival facilities. Rightly in 2010, the Chinese government launched the *Construction Planning for County-level General Archives in Central and Western China* ("the Planning") to grant central government subsidy to help the new construction and reconstruction of those undersized general archives at the county level, so as to improve the relatively backward conditions of the general archives at the grassroots level. As of 2019, the central government has allocated a subsidy amounting to CNY 6.9 billion in constructing more than 1,500 county-level archival facilities. As the new archival buildings listed in the Planning are completed and put into operation, the central government investment has yielded a remarkable result. The long-existent problem of"insufficient building" and "dilapidated building" found in county-level general archives in central and western China, are completely defused; the security risks against archives are eliminated as new repositories are installed with security and fire control facilities. These fundamental improvements of archival facilities help to take over by law and keep safe the archives from all sectors of the society.

Compared to the old ones, the newly built archival facilities are supposed to take on more social functions, and therefore have a more reasonable internal layout to accommodate more friendly exhibition halls, more comfortable file view zones, etc. Some adopt a more humanized design to create a more intimate and unaffected space, such as the connotation development of atrium and the building of leisure spaces inside and outside the building. With favorable hardware conditions, archival facilities nowadays can provide abundant archival services for the government and the public, including more convenient archival utilization, meaningful archival culture exhibitions, to strengthen the association between governments at all levels with the mass and to assist in the implementation of pro-people policies. These new buildings also have more distinctive appearances than the old ones. Some are designed with local historical and cultural characteristics to live in harmony with the existing structures around, and some others incorporate archival-specific design elements in new spaces to create eye-attracting landscapes for the city. To a certain extent, these buildings standing still in the center of cities get the archival work known by more people.

On the occasion of the 70th anniversary of the founding of New China, the National Archives Administration of China picks out 70 excellent archival facilities and 70 representative archival facilities from the county-level general archives in central and western China built in recent years with central government subsidy, and make them into a book, as a tribute to the brilliant achievements made in the archival cause of China, and to the 70th birthday of our motherland. Looking back to the past, archival work has never been so penetrating into every sector and every link of economic society, and so influencing in the material and spiritual life of the broad masses. We believe in a growing status and role of archives and archival work ahead. The work of archival facilities will also rise to new heights on its way to legal management, opening up, and modernization.

目录 Contents

江西省 Jiangxi Province

Henan Province

Hubei Province

Hunan Province

广西壮族自治区 Guangxi Zhuang Autonomous Region

Hainan Province

Chongqing Municipality

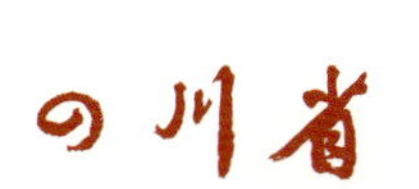

Sichuan Province

Guizhou Province

云南省 Yunnan Province

中西部地区县级综合档案馆建设规划

（国家发展和改革委 国家档案局 2010 年印发）

(Released by National Development and Reform Commission and National Archives Administration of China in 2010)

Construction Planning for County-level General Archives in Central and Western China

档案馆是集中管理档案的机构，负责收集、接收、整理、保管和提供利用各分管范围内的档案。作为档案事业的主体，它承担着保管人类社会各项活动真实记录的历史重任，以其丰富的馆藏档案，在服务经济建设、促进社会文化事业发展、维护国家利益与社会和谐稳定、保障群众权益等方面发挥了重要作用。为解决中西部地区县级综合档案馆馆舍面积严重不足的问题，增强档案馆库抗灾能力，保障国家档案资源安全，促进档案馆事业与本地区各项事业同步、协调发展，依据《中华人民共和国档案法》《〈中华人民共和国档案法〉实施办法》《档案馆建设标准》(建标 103—2008)《档案馆建筑设计规范》JGJ 25—2000 和《全国档案事业发展“十一五”规划》，在总结我国档案馆建设和发展经验的基础上，制定《中西部地区县级综合档案馆建设规划》(以下简称《规划》)。

本《规划》中“县级综合档案馆”是指按县级行政区划设置的、收集和管理所辖范围内多种门类档案的档案馆。

Archives are serving as organizations for centralized management of archive files to access, collect, sort, preserve and provide for access and use in their respective charge. Being as the subject of archival business, archives bear the important historical mission of preserving the true records of various activities of human society. With abundant collection of archive files, archives play a critical role in serving economic construction, promoting social and cultural development, safeguarding national interests and social harmony and stability, and protecting rights and interests of the public. For the sake of solving serious insufficiency of county-level general archives buildings in central and western China, promoting disaster-resisting performance of archives, safeguarding national archival resources, and upgrading synchronous and coordinated development of archive business with other undertakings in the regions, on the bases of archives construction and development experiences in China, the *Construction Planning for County-level General Archives in Central and Western China* (hereinafter referred to as “the Planning”) is hereby formulated in line with *Archives Law of the People's Republic of China, Implementation Measures of the Archives Law of the People's Republic of China*, *Construction Standard for Archives Buildings* (JB 103-2008), *Code for Design of Archives Buildings* (JGJ 25-2000) and *The National Archival Development Planning of the 11th Five-Year Plan Period.*

In the Planning, the “county-level general archives” are defined as per county-level administrative division, and take charge of collecting and managing various categories and classifications of archive files within their respective jurisdictions.

第一章

我国县级综合档案馆发展现状

在党中央、国务院及各级党委、政府的关心和支持下，全国档案馆事业得到了较快发展，档案馆网络已经基本形成。截至 2008 年，全国共有各级综合档案馆 3170 个，保存档案资料近 2.3 亿卷册。其中，县级综合档案馆馆藏档案数量 1.4 亿卷册。2008 年全国各级档案馆共接待利用者 2674771 人次，提供档案资料 9179012 卷（件），接待参观者达 1657836 人次。档案工作深入到了社会的各个方面，产生了深远影响，在我国政治、经济、文化和社会建设中发挥着越来越重要的作用。随着经济和社会的发展，社会各界对档案馆工作提出了新的更高要求，档案馆正在由之前单纯的档案保管、利用场所向档案安全保管基地、爱国主义教育基地、档案信息资源利用中心、政府信息查阅中心、电子文件中心“五位一体”的新型档案馆发展，逐步建立覆盖人民群众的档案资源体系、方便人民群众的档案利用体系和确保档案安全保密的档案安全体系。同时，档案馆的建设也面临着巨大机遇和挑战。

2018 年
全国共有各级综合档案馆
3315 个
保存档案资料近
78934.2 万卷（件、册）
全国各级档案馆共接待利用者
773.4 万人次
提供档案资料
1909.8 万卷（件、册）次
接待参观者达
486 万人次

在我国档案事业体系中，县级综合档案馆保管档案数量较大，服务公众群体较为广泛，与经济社会活动联系密切。由于我国幅员辽阔，各地经济社会发展差异较大，中西部地区县级综合档案馆发展还面临着诸多问题和困难，档案馆馆舍面积不足等问题已经成为制约档案事业发展的瓶颈。主要表现在：

一、中西部地区县级综合档案馆馆舍面积不足，难以保证档案正常接收，大量档案面临损毁的危险。中西部地区县级综合档案馆绝大多数建于 20 世纪 80 年代之前，档案馆库容已严重饱和，无法正常履行《中华人民共和国档案法》规定的依法接收档案的职能。拟列入本《规划》的县级综合档案馆现有平均建筑面积 501m^2，其中库房面积 224m^2。拟列入规划的档案馆馆藏档案实际存档量为 99826882 卷（件、册），受档案馆馆舍面积狭小的制约，到期未进馆档案数量达 36755075 卷（件、册），占现有馆藏总量的 27%。由于馆舍面积狭小，档案馆技术用房和展览利用用房短缺，造成了档案馆无法满足《中华人民共和国政府信息公开条例》（以下简称《政府信息公开条例》）所赋予的功能要求。

二、中西部地区县级综合档案馆建筑不符合《档案馆建筑设计规范》JGJ25-2000 要求，档案的安全难以保证。据各地上报数据显示，中西部地区县级综合档案馆现有 400 多个危房馆，大部分档案库房的设计载荷不达标；部分档案馆存在较为严重的消防安全隐患；绝大部分档案馆达不到《档案馆建筑设计规范》的要求，馆藏档案破损严重，虫蛀、鼠咬、受潮发霉、字迹褪变等现象较为普遍。

三、中西部地县级综合档案馆设施设备的配置严重不足，现有设施设备老化损坏严重，档案的防虫、防霉与复制等保护工作均无法正常开展。中西部地区县级综合档案馆具有大专以上学历的工作人员不足人员总数的 70%，经过档案专业培训的人员仅占人员总数的 35%，档案馆人员专业水平相对较低。这些问题都严重制约了档案馆工作的开展。

出现上述问题的主要原因包括：一是地方政府对档案工作的重要性认识不足，落实国家相关政策积极性不够；二是部分地方政府受财力的制约，对档案馆的发展长期投入不足，档案馆建设标准较低；三是档案的管理模式和管理体制有待进一步创新、完善，吸引和稳定人才的有效机制尚需进一步完善。

Chapter One

Current Situations of County-level General Archives in China

Under close care and support of the Chinese Party Central Committee, the State Council, and party committees and governments at different levels, the archives business in China have developed rapidly, with a national network of archives being basically taken shape. As of 2008, there were 3,170 general archives of all levels throughout the country, containing nearly 230 million files and volumes in their collections, of which 140 million preserved by county-level general archives. In 2008, archives of all levels around China received 2,674,771 persons, provided 9,179,012 archive files (items), and received 1,657,836 visitors. Archiving has involved in all sectors of the society with far-reaching influences, and is playing an increasingly important role in the political, economic, cultural and social development of China. As economy and society grow, all sectors have put forward new and higher requirements for the work of archives, which, adapting to this trend, are transforming from the traditional single role as places of archival storages and access toward a "five-in-one" functional body, i.e. a place functioning as a base of preserving archive files, a base of patriotism education, a center of archive information and resources utilization, a center of government information reference, and a center of electronic files. Thus an archival resources system covering the public will be gradually established, as well as an archival utilization system easily accessible to the public and a secured, confidential archival security system. To this end, there are huge opportunities and challenges to archives construction.

2018

China has a total of **3,315** general archives at all levels, with a collection of **789,342,000** archival files (items and volumes)

In 2018, the archives in China have provided **19,098,000** archival files (items and volumes) for the access and use of **7,734,000** person-times in total as well as **4,860,000** person-times of visitors

In China's archiving system, county-level general archives have huge amount of archives in custody, which serve for a large population and have close relations with economic and social activities. Due to China's vast territory and large differences in economic and social development in different regions and cities, the development of county-level general archives in central and western regions still faces many obstacles. The inadequacy of archives floor areas has become a bottleneck restricting the development of archives business showing in the following aspects:

Ⅰ. The county-level general archives in central and western China are inadequate in areas, which make it difficult to receive normal amount of archive files and a large number of archive files are being damaged. Most of county-level general archives in central and western China were built before the 1980s, with their repository capacities being overused and their accession function fails to fulfill the provisions of *Archives Law of People's Republic of China*. The county-level general archives to be included in the *Planning* have an average floor area of $501m^2$, of which the repository area is $224m^2$. The actual archiving capacity of the county-level general archives to be included in the Planning is 99,826,882 files (items/volumes). Constrained by the small sizes of the archives, the number of archive files that have not been admitted into the archives on due time has reached 36,755,075 files (items/volumes), accounting for 27% of the total collection. For the same reason, rooms for technical and exhibiting uses in the archives are insufficient, which has made them unable to perform functional requirements set forth in the *Open Government Information Regulations of the People's Republic of China* (hereinafter referred to as the Open Government Information Regulations).

Ⅱ. Since the general archives buildings at county level in central and western China do not conform to the requirements of *Code for Design of Archives Buildings*, it is difficult to secure the archive files therein. According to the data reported from local governments, more than 400 among the county-level general archives in central and western China have dangerous houses, and most of the archival repositories are not qualified with their design loads; part of archives buildings have serious fire risks; most archives buildings are not up to the requirements of *Code for Design of Archives Buildings* (JG25—2000), where serious damage, insect and rat bites, moisture, mildew and fading of writing are commonly occurring.

Ⅲ. Facilities and equipment in county-level general archives in central and western China are inadequate. Existing facilities and equipment are seriously damaged, which makes it impossible to properly take protective measures against insects, mildew and reproduction of archive files. Less than 70% of the total staff of county-level general archives in central and western China have received college education or above, and only 35% of the total staff have received professional training of archive science. Due to relatively low professional competence of archive staff, it seriously hampers the archives to function properly.

Main reasons for explaining the above problems include: first, local governments pay less importance to the archive business and less reaction to implement the national policies concerned. second, constrained by financial support, some local governments have insufficient input in long development of archives buildings, and the construction standards of archives buildings are relatively low. third, the management mode and system of the archives need to be further innovated and improved, and the mechanism for attracting and stabilizing talents needs to be further improved.

第二章

指导思想、建设目标和建设原则

第一节　指导思想

以邓小平理论和“三个代表”重要思想为指导，深入贯彻落实科学发展观，以为党和国家中心工作及各项建设事业提供有效服务为目标，以档案基础设施建设为核心，以档案安全保管保护为重点，努力提升中西部地区县级综合档案馆的整体服务能力和水平，促进档案馆事业与经济社会协调发展。

第二节　建设目标

通过中央和地方政府共同投入，加快中西部地区县级综合档案馆建设，逐步使现有档案馆舍达到《档案馆建设标准》和《档案馆建筑设计规范》的要求，改善中西部地区县级综合档案馆建设严重滞后的状况，解决长期以来档案不能按时接收进馆的难题，使档案馆能够正常履行《中华人民共和国档案法》《政府信息公开条例》所赋予的职能，档案的安全保管和有效利用得到可靠保证，档案抢救和保护的能力得到一定提升，为档案馆实现“五位一体”功能打下坚实基础。

第三节　建设原则

一、统一规划，分级负责。中央负责编制、审核中西部地区县级综合档案馆建设总体规划，适当安排补助投资，督导检查规划实施情况。各省（自治区、直辖市）以及市（地、州、盟）负责监督、协调、指导本地区建设项目的实施，安排补助资金，确保建设用地和建设资金的及时足额到位；县（市、区、旗）负责提供档案馆的建设用地，落实建设资金，减免有关税费，并负责项目的具体实施。

二、科学合理，兼顾发展。档案馆建设要严格遵循国家相关法律法规，按照《档案馆建筑设计规范》的要求，因地制宜、安全实用、简洁朴素、高效节能，不得贪大求洋、超标准装饰装修，满足档案馆“五位一体”的功能设置要求及档案事业长远发展的需求。

三、合理安排，配套推进。各地在编制年度中央投资计划时，应区分轻重缓急，优先考虑无馆舍、危房馆舍和问题严重、条件艰苦的档案馆。在加强基础设施建设的同时，各地要加快建立档案馆运行保障机制和财政保障机制，配备档案馆运行必需的设施设备，加强人才培养，提高服务水平。

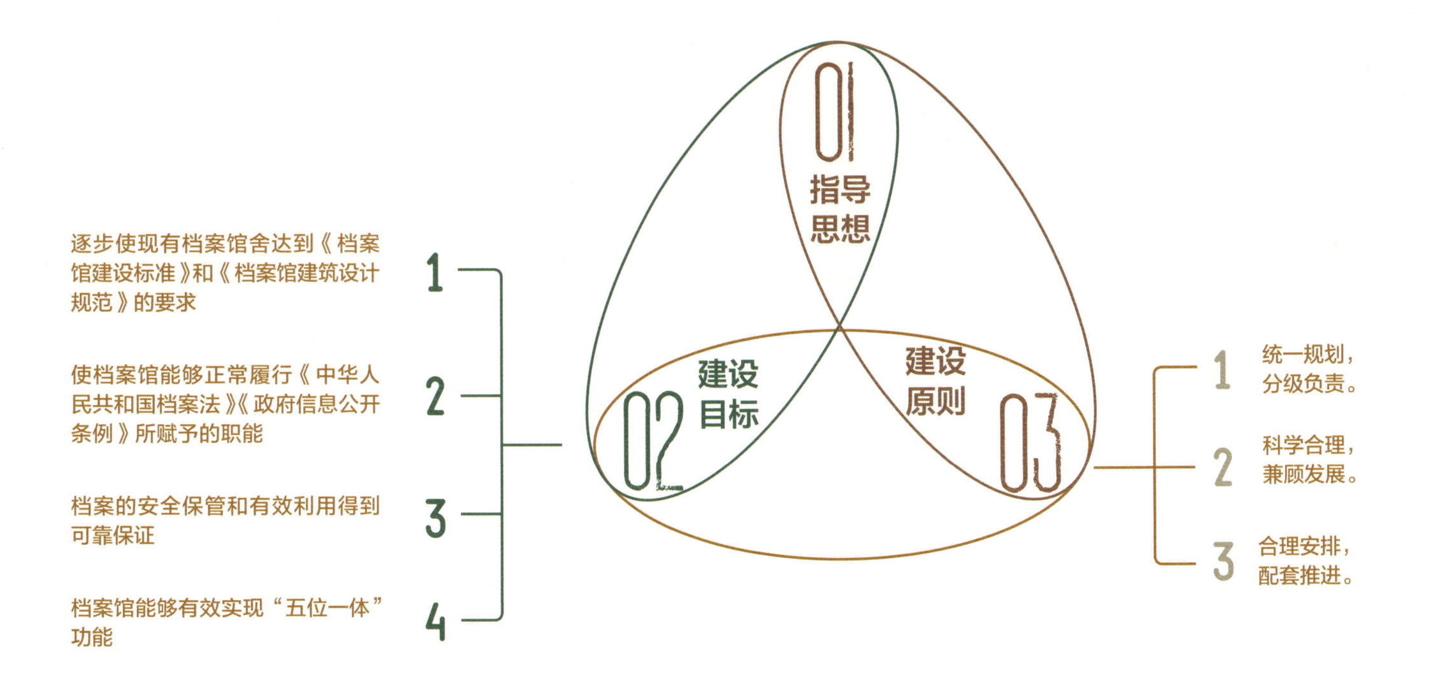

Chapter Two

Guiding Ideology, Construction Objectives and Construction Principle

Section 1 Guiding Ideology

Guided by "Deng Xiaoping Theory" and the critical thoughts of the "Three Represents", the Planning implements the scientific development concepts with the aim to provide effective service for the major tasks of the Party and the state, with the core mission of construction of archival infrastructures and preserving and safeguarding archive files, so as to enhance the overall service competence of county-level general archives in central and western China, and promote coordinated development of archiving, economy and society.

Section 2 Construction Objectives

With joint investments of the central and local governments, the construction of county-level general archives in central and western China is expected to be accelerated, and the existing archives buildings to reach the requirements of *Construction Standards for Archives* and *Code for Design of Archives Buildings*, construction of county-level general archives in central and western China which is seriously lagged behind will be improved, and the problem of failing to receive and access archive files on due time will be solved. In this way, archives can properly perform their functions assigned by *Archives Law of People's Republic of China* and the *Open Government Information Regulation*; the secured storage and effective access of archives can be reliably guaranteed; and the performance of recovering and conserving the archives can be improved, laying a solid foundation for the archives to assume the "five in one" role.

Section 3 Construction Principles

Ⅰ. To make general planning and define responsibilities in tiers. The central government is responsible for formulating and reviewing the general planning for construction of archives in central and western China, appropriately allocating investments, and supervising over implementation of the general planning. All provinces (autonomous regions and municipalities directly under the central government) and cities (prefectures, states and leagues) should be responsible for supervising, coordinating and guiding implementation of local construction of the archives, and arranging the investments to ensure construction lands and investments are ready on due time in proper amount; counties (cities, districts and banners) are responsible for providing construction lands for the archives, using investments on due purpose, reducing and exempting relevant taxes, and responsible for specific implementation of construction.

Ⅱ. To be scientifically reasonable, taking development into account. The construction of archives should strictly abide by relevant national laws and regulations, conform to the requirements of *Code for Design of Archives Buildings*, being adaptable, safe, practical, simple, efficient and energy-saving, and avoid pursuing big size, unsuitable styles or making excessive decorations. Functional requirements of archives on the "five in one" principle and the demand of long-term development of should be met.

Ⅲ. Making reasonable arrangements with proper support for construction of the archives. When preparing the annual central government investment plans, all local governments should prioritize the urgent and important issues, first addressing the places where archives buildings are absent or archives buildings in poor or dangerous conditions. While infrastructures are reinforced, all local governments should quickly establish an archives operation and financial safeguarding mechanisms, furnish the archives with necessary facilities and equipment, and strengthen talent training to improve overall service quality.

第三章

建设计划

档案馆建筑由档案库房、对外服务用房、档案业务和技术用房等主要功能用房和附属用房组成，档案库房是档案馆履行基本职能的核心。

县级综合档案馆的建设规模主要以应保存的馆藏档案数量为基本依据，即：馆藏在10万卷以下的建筑面积在1200～2600m^2，馆藏10万～20万卷的建筑面积在2600～4600m^2，馆藏20万卷以上的建筑面积在4600～6800m^2。具体用房安排和设施设备配置，原则上按照《档案馆建设标准》执行。

应保存的馆藏档案数量是指现有馆藏档案数量和今后30年依据《档案法》的规定应进馆的档案数量之和。现有馆藏档案数量、现有馆舍面积的测算核定依据是2005年《全国档案事业统计综合年报》，年均档案接收数量测算的依据是连续十年《全国档案事业统计综合年报》年档案接收数量的平均值。对部分因库容饱和无法正常接收档案的县级档案馆，其年均接收档案数量参照本地区人口规模相近的其他县级档案馆的接收数据。

档案馆建设的选址和总平面、建筑设计、档案防护、防火和建筑设备必须符合《档案馆建筑设计规范》等规范、标准的要求。选址要充分考虑到方便人民群众和社会对档案利用的需求。

中西部地区县级综合档案馆建设于2010年启动。各地要依据本《规划》精神，根据项目建设需要落实建设资金和各项条件，加大投入力度和监管力度，确保按期完成建设任务。

各地要严格按照国家有关规定，履行建设程序，认真执行国家有关建设标准与规范，不得压缩经核定的档案馆建设规模，有条件的地区可结合实际，适当增加建设规模。

要加强对项目和资金的监督与管理。中央预算内专项投资及地方配套资金，要做到专款专用。项目完成后，各省统一组织验收。

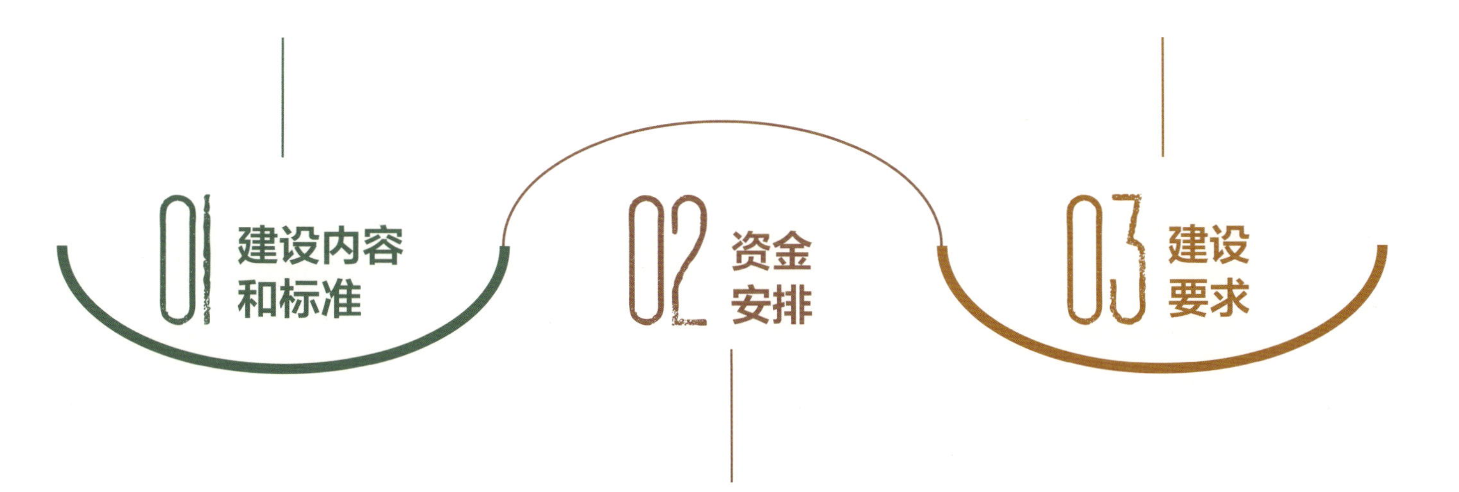

一、中西部地区县级综合档案馆建设投资由中央和地方共同承担，各级政府要切实加大对档案馆建设的投入力度。中央投资的补助标准和补助额度，在安排年度计划时，结合物价水平等因素确定。“十二五”期间，中央投资主要支持库房面积小于200m^2（不含）的档案馆项目建设。

二、中央补助投资支持的重点是：西部地区和享受西部待遇的中部地区档案库房、对外服务用房、档案业务用房等馆舍建设的建安工程费用（含改造费用）；中部地区档案馆档案库房建安工程费用（含改造费用）。

三、除中央补助投资外，其余建设资金及建设规费由地方政府筹措解决。对于县级人民政府在资金配套上确有困难的，由省、市级人民政府统筹解决。档案馆建设所需的建设用地，由地方政府无偿提供。

四、档案馆建设属地方事权。国家鼓励地方政府加大对档案馆建设的投入力度。库房面积大于等于200m^{2}的县级综合档案馆建设，由各地方政府视地方财力，积极安排建设资金并组织实施，确保档案馆建筑达到《档案馆建设标准》和《档案馆建筑设计规范》的要求。

Chapter Three

Construction Plan

Section 1 Construction Contents and Standards

An archives building should consist of main functional rooms such as archival repositories, public service rooms, archives business and technical rooms, as well as auxiliary buildings. Archival repository is the core for archives to perform basic functions.

The construction scale of a county-level general archives is mainly based on the number of archives to be collected, that is, the floor area should be 1,200-2,600m^2 for archives with a collection less than 100,000 files, 2,600-4,600m^2 for archives with a collection of 100,000-200,000 files, and 4,600-6,800m^2 for archives with a collection more than 200,000 files. Specific allocation of rooms, facilities and equipment should conform to *Construction Standards for Archives*.

The "number of collection" refers to the number of existing collection and the number of archives accession in the next 30 years pursuant to *Archives Law*. The estimation and verification of the number of existing collections and the area of existing buildings should base on *Annual Report of National Archives Statistics 2005*. The "annual average number of archives accession" is based on the average of archives accession in the *Annual Report of National Archives Statistics* for ten consecutive years. For those that cannot receive archives normally as they have reached full capacity, the average number of archives accession per year should reference to the archives accession of other archives in cities with similar population in the region.

The site selection, master plan, architectural design, archive conservation, fire prevention and building equipment in archives construction should conform to *Code for Design of Archives Buildings* and other criteria and standards. The site selection should take into account demands of the public and the society.

Section 2 Investment Allocation

I. The investments in construction of county-level general archives in central and western China should be jointly assumed by the central and local governments, and governments at all levels should effectively increase inputs in archives construction. When an annual plan is being prepared, the subsidy standard and quotas of investment from the central government should take into account the price level and other related factors. During the "Twelfth Five-Year Plan" period, the central government will mainly put its investment in archives with repository area less than 200m^2 (excluding).

II. The focus of the central government to grant subsidies is: the costs for construction and installations (including renovation costs) to build repositories, public service rooms, archives business operation rooms in the western regions as well as the central regions enjoying the preferential policies as the western regions; the costs for construction and installations (including renovation costs) to build archival repositories in the central regions.

III. Other than the subsidies from the central government, the remaining construction investments and regulatory fees should be raised by local governments. In case the county-level governments have difficulties in raising the funds, such funds should be raised by provincial and municipal people's governments. The lands required for building archives should be offered for free by local governments.

IV. Construction of archives buildings falls in the power of local governments. The central government encourages local governments to increase inputs in archives construction. For construction of county-level general archives with repository area more than or equal to 200m^2, local governments should actively arrange construction funds and organize construction based on the local financial capacity to ensure that archives buildings to be built meet the requirements of *Construction Standards for Archives and Code for Design of Archives Buildings*.

Section 3 Construction Requirements

The program of building county-level general archives in central and western China was launched in 2010. All local governments should, in accordance with the spirit of the Planning, raise construction funds and prepare various conditions in accordance with the construction demands, increase inputs and supervision, and ensure that all construction tasks be completed on schedule.

All local governments should strictly follow relevant national regulations and construction procedures, earnestly implement relevant national construction standards and criteria, and should not shrink the approved construction scale of the Archives. Some regions can properly increase the scale of construction if applicable based on the actual local conditions.

It is necessary to strengthen supervision and management over construction projects and investments. Special investments granted by the central government and funds offered by local governments should be earmarked for the specific purposes. Upon completion archives projects should be inspected and accepted by the respective provincial government.

第四章

其他要求

为保障中西部地区县级综合档案馆项目的顺利完成，实现建设目标，各地档案馆建设必须遵循档案事业发展的客观规律，坚持加大投入与改革发展，坚持继承与创新，突出抓好馆藏建设，进一步加强公共服务能力建设，提高档案安全保障能力，完善投入保障机制，加快人才培养。

馆藏建设是档案馆事业发展的核心，要按照建立覆盖人民群众的档案资源体系的要求，下大力气抓好馆藏建设。各级综合档案馆要及时修改调整档案馆的接收范围，在确保基本馆藏的前提下，结合自身能力，将其他类别的档案及时接收进馆，做到应收尽收。要积极以多种形式进行档案资源整合。通过开展馆际协作及馆室之间协作等多种形式，结合本地区实际，不断创新本地区档案资源整合的模式，为档案资源社会共享打下基础。要通过购买、交换等方式，继续加快对散存在民间及境外的、具有保存价值的重要档案的征集步伐。鼓励个人向国家档案馆捐赠或寄存档案。

为全社会、为广大人民群众提供便捷的档案信息，是各级国家档案馆履行政府公共服务职能的重要任务。各级国家档案馆要紧紧围绕党和国家工作大局，围绕各级党委、政府的中心工作，围绕党政机关需要、社会需要和群众需要，利用各种手段和形式，及时提供档案资料，广泛开发档案信息资源，从各方面加强档案馆的公共服务能力建设，建立方便人民群众的档案利用体系，实现档案信息资源社会共享。

要严格按照相关规定和本《规划》的要求，建设各类档案的阅览场所、政府信息查阅场所、档案展览陈列场所、档案宣传教育场所、档案远程利用场所等，同时配置好相应的设备，使档案馆的场所、设备等硬件设施能满足各项功能和公共服务的需要。

要加大档案开发和提高利用力度，加快档案的整理。按照《中华人民共和国档案法》的规定，加大馆藏档案自形成之日起满 30 年档案的鉴定和开放力度，尽快做到应开放的档案资料全部开放。进一步做好政府信息查阅工作，完善政府信息的报送、利用等各项规章制度建设，为公民、法人或其他组织获取政府信息提供便利，把《政府信息公开条例》赋予各级国家档案馆的这一新职责切实履行好。

始终不懈地抓好安全建设，确保档案信息安全保密，提高档案安全保障能力，是各级综合档案馆的基本职责。档案馆库房是档案馆建筑的核心。汶川地震等经验教训告诉我们，合格的档案馆库可以在灾害发生时最大限度地保护档案安全，减少档案损失。档案馆库建设要充分考虑环境因素，采用防火、防水、无污染的建筑材料，在设计上减少强光、尘土、高温高湿、有害气体等对档案的危害。所有档案库房必须配备消防设施。温湿度不利于档案安全的地区的档案库房，还应安装温湿度调控设施。档案馆在设计阶段要考虑多种载体档案保护以及对档案原件的监控问题。各级档案部门要积极采用各种先进的安全防范技术手段，确保档案安全保密。

各级政府要高度重视县级综合档案馆建设和运行，充分考虑档案馆舍后续运行维修和管理工作，建立起档案馆后续维护的投入机制和管理机制，做好档案馆所需设备购置费用、档案保护费用以及日常经费等的落实工作，确保县级综合档案馆的正常运转。

提高档案干部队伍的综合素质是档案馆事业发展的关键。要培养出更多刘义权式的坚守平凡创造非凡、甘于吃苦乐于奉献、爱岗敬业执着追求的档案工作者；要加强档案部门的领导班子建设和干部队伍建设，从政治和业务两个方面教育和培养干部，通过各种方法，锻炼、培养领导班子和广大干部的大局意识、政治意识、群众意识、社会意识、服务意识，扩充、增长档案工作者的专业知识、法律知识、相关知识、专业技能，增强每个人的创新能力、执行能力，使档案干部队伍成为一支政治强、业务精、作风好、能力强的高素质队伍，为档案馆的发展提供强有力的人才保障和智力支撑。

《规划》实施过程中，各地要将年度建设项目安排情况向社会公开，接受社会监督。项目主管部门要对建设项目进展情况、工程建设质量、设备采购招标、资金到位和使用等情况进行经常性检查并落实工作责任。国家发展改革委、国家档案局及相关部门将对中央投资项目实施情况进行专项检查，对因资金管理不善造成损失和浪费、工程质量出现问题等违法违纪行为，一经查实，将依法追究责任。

Chapter Four
Other Requirements

For successful completion of county-level general archives projects in central and western China and of construction objectives, objective laws and rules for developing the archives industry should be followed. Keep reform and increase investment, adhere to inheritance and innovation, highlight construction of archival repository and public service performance, improve safeguarding ability and talent cultivation.

Section 1 Highlight Construction of Archival Repository

Being the core of archives development, repository construction should meet the requirements of building archives resource system for the people. General archives at all levels should promptly define and modify scopes of archives collections based on their respective demands and capability. Under this premise, other classifications of archives can also be collected. Integrate archive resources by various means. Through different forms of cooperation, more innovative modes of resource integration should be made based on the local conditions. Besides, through purchases, exchanges and other means, critical and valuable files and records estranged in individuals and abroad should be acquired in a quick manner. Individuals are encouraged to donate or deposit their archives at national archives.

Section 2 Strengthen Performance of Public Service

It is an important mission of national archives at different levels to conveniently provide archival information for the whole society and the public. National archives at all levels should closely focus on the main missions of the Party and the country, and on the demands of the government, society and people. Duly provide necessary and profound archival information through all means, strengthen overall performance of public service, establish an archive access and utilization system that is convenient for the people, and share archives information resources in the society.

It is necessary to strictly follow the requirements of relevant regulations and the Planning to build venues for accessing archives, inquiring government information, displaying archival files, publicizing information about archives, and offering remote access for archives. At the same time, facilitate all archives with related equipment so that the sites, their facilities and other hardware can fulfill all functions and public services.

It is necessary to make efforts to develop archives and enhance access utilities and speed up sorting and compilation of archives. In accordance with the provisions of *Archives Law of the People's Republic of China*, efforts should be made to identify and open to the public the archives that have been kept for 30 years or more since the establishment of an archival file, and open the access of all archives resources to the public as soon as possible. Further improve the work to inquire government information, improve all rules and regulations for reporting and utilizing government information, and offer convenience access for citizens, legal persons or other organizations to acquire government information, and truly perform obligations endowed by *Open Government Information Regulation* to national archives at all levels.

Section 3 Improving Performance of Safeguarding Archives

It is essential obligations of general archives at all levels to consistently keep confidential of archival information and improve performance of safeguarding archives. Archival repository is the core of an archives building. Lessons from the Wenchuan Earthquake tell us that qualified archival repository can maximize safeguarding archival information and reduce archive losses in disasters. The construction of archival repositories should take environmental factors into full consideration, and fireproof, waterproof and pollution-free building materials should be used to reduce harm to archives due to glare, dust, high temperature and high humidity, harmful gases, etc. All archives buildings should be equipped with fire control facilities. For those archival repositories where temperature and humidity are unfavorable for archives security, temperature and humidity control facilities should be otherwise installed. In the design stage of an archives building, archives conservation and monitoring approaches should be taken into consideration. Archiving authorities at all levels should actively adopt advanced safeguarding and technical measures to keep archives secured and confidential.

Section 4 Establishing Necessary Investment Guaranteeing Mechanisms

Governments at all levels should attach great importance to construction and operation of county-level general archives, fully consider their subsequent operation and maintenance, establish investment and management mechanism for subsequent maintenance, allocate funds for equipment purchasing, archives safeguarding costs and normal operation costs that are necessary for the archives, and safeguard normal operation of county-level general archives.

Section 5 Speeding up Talent Training

Improve overall quality and qualification of archival staff is the key for development of the archives business. Cultivate more archivists who are willing to devote themselves to the career, like Liu Yiquan who is doing extraordinary work on his excellent post; more efforts should be made to build management teams, offering both political and business training programs. The management teams should be highly aware of political, public, social and service issues. They should expand and increase their professional skills and knowledge. They should have stronger performance in innovation and execution, and higher strength in political awareness and expertise. A strong team of archival leaders and staff is a solid foundation for archives development.

Section 6 Implementation Supervision Mechanism

In the course of implementing the Plan, all local governments should make it open to the public about the annual construction projects of archives buildings so as to be supervised by the society. Competent authorities of projects should make regular inspections on progresses, quality, biddings for equipment procurement, and availability and use of investments. The National Development and Reform Commission, the National Archives Bureau and relevant authorities should conduct special inspections on the projects funded by the central government. In case any violation against laws and regulations is observed, such as loss, waste or quality issue due to poor management of funds will be strictly investigated and punished.

第五章

预期建设成效

在中央和地方政府的共同努力下，逐步改善中西部地区县级综合档案馆馆舍建设严重滞后的现状，力争解决长期存在的无馆舍、有馆无库、危房、档案保管条件恶劣、因库房狭小档案不能按期进馆等严重制约档案馆发展的根本性问题，实现档案依法接收和国家档案资源的安全保管，使中西部地区县级综合档案馆基本达到《档案馆建设标准》的要求，真正成为档案安全保管基地、爱国主义教育基地、档案信息资源利用中心、政府信息查阅中心、电子文件中心。县级综合档案馆将更好地服务民生、服务人民群众、服务党和国家中心工作，也必将在促进经济建设、政治建设、文化建设、社会建设，在推动科技进步、提高公民素质、维护祖国统一、增强民族团结、维护社会稳定等各项工作中发挥更大的作用。

Chapter Five

Prospect of Construction Achievements

Through joint efforts of the central government and local governments at all levels, the lagged-behind situation of county-level general archives building construction in central and western China is expected to be improved. The long-standing, fundamental issues that severely restrict development of these archives are expected to be solved, such as insufficiency of archives buildings, insufficiency of repositories, buildings in dangerous and poor conditions, and inaccessibility of archival files, etc. Through joint efforts, archives accession can be realized in line with related laws and national archival resources can be conserved in safe custody. With all these measures, county-level general archives in central and western China are expected to meet requirements of *Construction Standards for Archives* and truly realize such functions as being bases of safeguarding archival resources, bases of patriotism education, centers of archival information and resources utilization, centers of government information, and centers of electronic files. County-level general archives will do a better job serving for the people's livelihood, the public, and the Party and government's core missions. They will also play a more significant role in promoting economic, political, cultural and social constructions, as well as in scientific and technological improvement, increasing citizens' quality, safeguarding national unity, consolidating ethnic solidarity and maintaining social stability.

中西部地区县级综合档案馆建设优秀实例

Demonstration Cases of Construction for the County-level General Archives in Central and Western China

献县档案馆

Xian County Archives

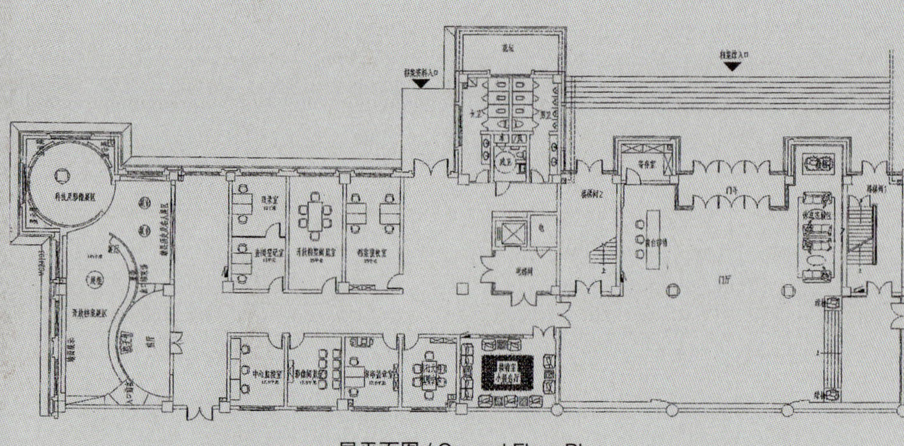

一层平面图 / Ground Floor Plan

占地面积： 10000m²
建筑面积： 4439m²
库房面积： 1484m²
层　　数： 4 层
设计单位： 浙江东华规划建筑园林设计有限公司
建成时间： 2015 年 12 月

Site area: 10,000m²
Floor area: 4,439m²
Repository area: 1,484m²
Number of floors: 4 floors
Time of completion: December 2015

献县档案馆始建于1956年，现有馆藏档案117个全宗，馆藏总量7.95万卷（件），资料2.3万册。新馆于2014年12月正式开工建设，2015年12月主体工程全面竣工。

新建的档案馆与博物馆、图书馆、规划馆、文化馆“五馆合一”，集中于一栋建筑中，作为献县城市文化中心。建筑平面为长方形，有两个内置庭院，形如“日”字。各馆分区明确又相互联系，均有独立入口，档案馆位于北侧。

档案馆采用框架结构，地上4层。一层设有门厅、展厅、查阅登记室、目录室、阅览室、复印室等对外开放服务功能，其中服务大厅与展厅为一、二层通高空间，营造出开敞开阔的公共氛围；二、三、四层东侧房间进深较大，作为档案库房区；西侧靠近内院、进深较小的空间采取中间走廊南北两侧布置房间的方式，作为档案业务、技术用房以及办公用房，拥有较好的通风、采光与景观视野。

档案馆库房设施设备按照智慧档案馆和全国示范数字档案馆的标准进行设计，安装智能密集架1500m³，其中包括音像防磁柜、底图柜、图书报刊架、实物油画架等异形密集架。库房环境配备了中央空调、恒温恒湿一体机、门禁视频监控系统，气体灭火系统。全部库房由一体化智能管理平台系统进行管理。

档案馆新馆结合献县献王文化特色，采用中国传统坡屋顶，出挑深远，水平延展，呈现出新中式汉风的建筑形象。立面为主入口突出两侧对称的三段式做法，以竖向的条窗与柱廊进行划分，虚实结合，韵律感与节奏感十足。

整个建筑宏观上追求城市文化建筑的庄重、大气，一气呵成；微观上对建筑细部精雕细刻，仿石材真石漆与玻璃交相辉映，简约而不简单，简洁而富有内涵。

1 档案馆外景 / The Archives Exterior

The building of Xian County Archives was constructed as early as 1956, The Archives has 117 fonds, with a total collection of 79,500 files (items) of archives and 23,000 volumes of documents.

The construction of the new Archives was officially commenced in December 2014, and the main building was completed in December 2015.

The newly built Archives and the county's Museum, Library, Planning Hall and Cultural Center were "integrated into one", which means they were all put into the same building to become a cultural center of the county. The building is rectangular in shape with two courtyards inside, making it appear the Chinese character " 日 ". The five organizations are both independent of and interconnected with each other, with accesses from their respective entrances. The Archives is on the north side.

The Archives applies a framework structure with four stories aboveground. On 1F there are the Hallway, Exhibition Halls, Check-in Room, Catalogue Room, Reading Room, Photocopying Room and rooms for other public services. The Service Hall and Exhibition Hall are wide and open public spaces with full height from 1F to 2F. Rooms on the east side on 2F, 3F and 4F have large inmost depth, and are used as Archival Repositories. In the spaces that are close to the courtyard on the west, a corridor is arranged in the middle with rooms on the south and north sides, which are business rooms, technical rooms and offices, as they have favorable ventilation, lighting and line of sight.

The repositories are designed in accordance with the standards for intelligent archives and national demonstration digital archives. 1,500m^3 of intelligent compact shelving have been installed, including audio-visual anti-magnetic cabinets, base-drawing cabinets, books and newspaper shelving, oil painting shelving, and other irregular compact shelving. The repositories are equipped with central air conditioning, constant temperature and humidity machines, access control CCTV system and gas fire control system. All repositories are managed through an integrated intelligent management platform.

The cultural characteristics of King Xian of Xian County are fused in the new Archives' design via the adoption of traditional Chinese pitched roof with a far-reaching and horizontal extension, presenting a new architectural image of the Han style. The building's facade built in the three-piece style to give prominence to symmetry on both sides at the main entrance. Vertical strip-shaped windows and colonnades act as the separator to combine virtuality with reality and bring about strong sense of rhymes and rhythms.

Macroscopically, the building presents solemnness, magnificence and smoothness as an urban cultural architecture. Microscopically, it is a piece of work forged with care and precision. Artificial stones, stone-like coating and glasses complement each other, making it a simple, concise building with rich connotations.

2 | 3
4
5 | 6

2 一层大厅 / GF Lobby
3 库房 / Repository
4 展厅 1 / Exhibition Hall 1
5 展厅 2 / Exhibition Hall 2
6 展厅 3 / Exhibition Hall 3

高平市档案馆

Gaoping Archives

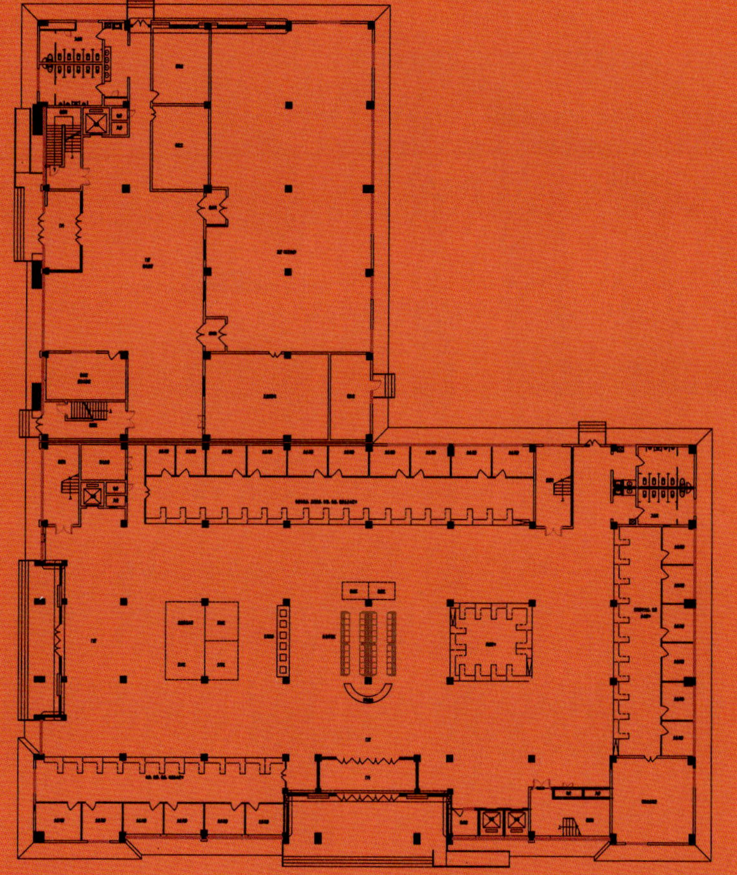

一层平面图 / Ground Floor Plan

占地面积：1242m²
建筑面积：6438m²
库房面积：2010m²
层　　数：地上 5 层，地下 1 层
设计单位：太原市筑博建筑设计有限公司
建成时间：2016 年 5 月

Site area: 1,242m²
Floor area: 6,438m²
Repository area: 2,010m²
Number of floors: ↑ 5 floors and ↓ 1 level
Time of completion: May 2016

高平市档案馆是 1963 年 4 月在高平县委办公室档案室的基础上成立的。现馆藏档案资料 58630 余卷（册），其中档案 42630 卷（涉及 106 个全宗 132 个机构），资料 16000 余册（卷、件）。

高平市档案馆新馆为合建项目，与合建的公共资源交易中心相互独立，自成一区。该项目西临“三馆三中心”，与高平市图书馆、博物馆、文化馆，老干部活动中心、青少年活动中心、妇女儿童活动中心形成建筑群，作为高平市城市文化中心，东临丹河、南临康乐街、北临炎帝公园，地理位置优越，周围环境良好。

档案馆建筑采用框架结构，抗震设防烈度 7 度，地上 5 层。主入口位于一层，面向中心广场，本层设有服务大厅、展厅、档案接收等对外开放功能；二～五层布局相似，均由档案库房区和围绕库房的功能性房间组成：二层为库房与开放阅览室，三层为库房与多功能厅、培训室、会议室，四层为库房与档案加工处理等技术用房，五层为库房与业务用房。整个建筑功能分区合理，使用方便。

馆内配置有先进的智能档案密集架、智能书车、细水雾喷淋和七氟丙烷气体灭火系统、智能监控防护系统，是全省一流水平的县级国家综合档案馆。

档案馆形体呈 L 形，面向河道打开，广场与城市道路一侧建筑界面完整。立面采用简洁现代的建筑语言，与西侧的“三馆三中心”协调统一。围合出的 U 形广场空间以开放的姿态拥抱着这座城市。

高平市档案馆新馆集档案安全保管基地、爱国主义教育基地、档案利用中心、政府信息查阅中心和电子文件中心为一体，符合国家县级一类档案馆设计规范和建设标准，满足未来 30 年档案进馆之需要。新馆的落成，有力地提升了档案工作的社会地位和法制意识，为档案工作服务于全市经济社会发展大局、加快“大美高平”建设、推动高平市档案工作的进一步发展，打下了坚实的物质基础。

1 档案馆外景 / The Archives Exterior

Gaoping Archives was established in April 1963 with the Archives Office of the Gaoping County Committee's General Office as its predecessor. The Archives has more than 58,630 volumes (copies) of archives in its collection, including 42,630 volumes (involving 106 fonds and 132 organizations) of archival holdings and more than 16,000 copies (volumes, pieces) of other documents.

The new site of Gaoping Archives is constructed jointly with the Public Resources Trading Center as a self-contained area. Being adjacent to the "Three Halls and Three Centers" in the west, the project is a part of a buildings cluster which also contain Gaoping city's Library, Museum, Cultural Center, Senior Officials' Day Center, Children's Day Center, Women and Children's Day Center; this cluster, as the urban cultural center of the city, is next to the Dan River on the east, to Kangle Street on the south and Yandi Park on the north with a superior location and nice ambient environment.

The Archives building adopts a framework structure that is able to proof earthquake with Level 7 seismic fortification and five floors above the ground. Facing the central square, the main entrance is located on 1F. On the same floor public functions are provided including the Service Hall, Exhibition Hall, Archives Accession, etc. 2F to 5F share the similar layout consisting of Archival Repositories and Repository Auxiliary Rooms: 2F accommodates a Repository and a Public Reading Room; 3F a Repository, a Multi-purpose Hall, Training Room and Conference Room; 4F a Repository, Archive Processing Room and other technical rooms; and 5F a Repository and Business Room. Functional areas in the entire building are reasonable arranged for convenient use.

Equipped with advanced intelligent compact archives shelving, smart book trolleys, mist sprinklers and heptafluoropropane gas fire extinguishing system, and intelligent monitoring and protection system, this county-level national comprehensive archives is among the top archives of the province.

The Archives is L-shaped in appearance and opens to the river, with complete architectural interfaces on the side facing the square and urban road. The facade is built in a simple, modern architectural language in coordination with the "Three Museums and Three Centers" on the west. The U-shaped square space in enclosure embraces the city in the open.

The new Archives, by providing an archives storage base, a patriotism education base, an archives access and use center, a government information center and an electronic document center, conforms to the design specifications and construction standards for national class-1 archives at the county level, and will fulfill the needs for the next 30 years. The new site of Gaoping Archives has effectively enhanced the social status and legal awareness of archiving, and laid a solid foundation for archiving to serve the overall economic and social development of the city, accelerate the construction of a *beautiful Gaoping*, and promote further development of the city's archiving business.

2 | 3
4
5

2 档案利用大厅 / Archival Access and Use Hall
3 中控室 / Central Control Room
4 档案培训中心 / Archival Training Center
5 库房 / Repository

Established in November 1958, Qin County Archives has a total collection of 150,000 files (items, volumes) of archives in six classifications, namely administrative, business, accounting, science and technology, audio and video, and material objects.

Located in the southwest corner of Nanhu Park of Qin County, the new site of the Archives is a modern national archives with a novel design, reasonable collection structure, advanced equipment, complete classifications and functions.

The new building has a simple, elegant U-shaped spatial layout. In the center of 1F is the entrance Service Hall. Exhibition Hall and Video Hall, which display the history and latest development of the county, are seated on the east and west wings respectively. In order that the public service functions will be offered in a better way, the archive also provides an Archives Reference Room, Public Reading Room, Prevailing Document Reference Room, Multimedia Lecture Hall, etc., which are located in the middle of the U-shaped space on 2F and 3F for public to easy access. Archives Repositories are on the east and west wings.

The repositories are divided into four areas, i.e. intelligent, material storage, compact shelving and archives backup. In all of the repositories, automatic temperature/humidity control system, automatic fire-extinguishing system and CCTV system, as well as AC system are available to improve the overall capacity for archive preservation.

Highlighting the new concept "facing the future and the public", Qin County Archives is both contemporary and local featured. It interprets the city's memory in a modern architectural language and decorative style, showing profound historical accumulation and unique culture of this river city in Northern China. Qin County Archives has become not only a repository of both state archives and social resources, but also an important platform serving both for the citizens and the society. It has accumulated valuable experiences for constructing city archives, and plays as a model for building county-level new archives throughout the province.

2
3 | 4
5 | 6

2 档案查阅大厅 / Archives Reference Hall
3 门厅 / Hallway
4 会议中心 / Conference Center
5 档案查阅室 / Archives Reference Room
6 庭院 / Courtyard

阳泉市郊区档案馆

Yangquan Suburb Archives

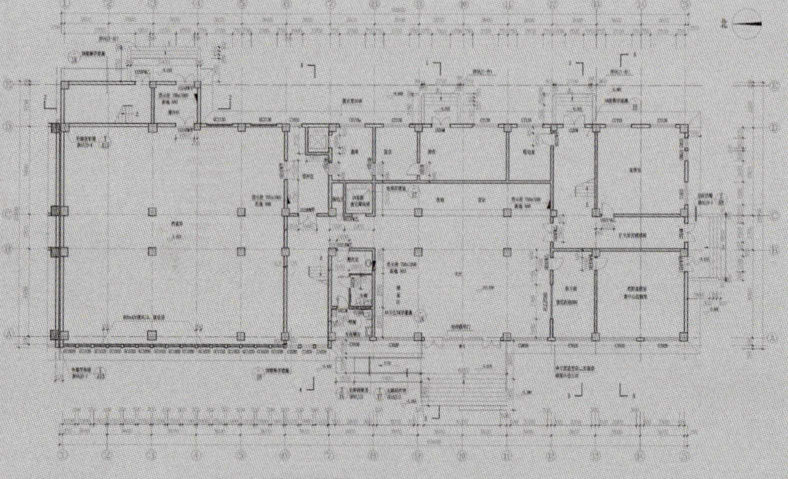

一层平面图 / Ground Floor Plan

占地面积： 3346m²
建筑面积： 5127m²
库房面积： 2082m²
层　　数： 主楼 6 层
设计单位： 山西省阳泉市建筑设计院
建成时间： 2012 年 12 月

Site area: 3,346m²
Floor area: 5,127m²
Repository area: 2,082m²
Number of floors: 6 floors for the main building
Time of completion: December 2012

阳泉市郊区档案馆始建于 1971 年，现有馆藏档案资料 124715 卷（件）。2013 年晋升为国家一级档案馆，2015 年被评为全国中小学档案教育社会实践基地。

阳泉市郊区档案馆新馆坐落于山西省阳泉市郊区荫营镇江正大街，于 2010 年 12 月开工建设，2012 年 12 月建成并投入使用。

新馆建筑采用几何形体穿插而成，简约现代。建筑布局合理，分为档案库房、公共服务区和办公区，三区在功能上既有效联系又相互独立，设有档案库房、多功能报告厅、展览厅、会议室、陈列室、数字化加工整理中心、消防监控中心、计算机管理中心、信息查阅利用中心、办公区等。

馆内配置了先进的气体自动灭火、自动喷淋、防火防盗自动报警、电子监控、中央空调、指纹识别门禁、净化消毒、加湿除湿一体机等智能化管理系统和现代化管理设施设备等。

阳泉市郊区档案馆已成为阳泉郊区新城的地标性建筑，是一座集档案资料保管基地、爱国主义教育基地、中小学档案教育社会实践基地、现行文件利用中心、政府信息公开查阅中心、电子文件中心和政务会议中心为一体的现代化公共档案馆，彰显了开放公共档案馆的建设理念。

1 档案馆外景 / The Archives Exterior

Initially built in 1971, Yangquan Suburb Archives has a collection of 124,715 files (items). In 2013, it was upgraded to a national class-1 archives. In 2015, it was elected a national social practice base for archiving education in primary and secondary schools.

The new site of Yangquan City Suburb Archives is located in Jiangzheng Street, Yinying Town in a suburb of Yangquan City, Shanxi Province. The construction was started in December 2010 and was completed and put into use in December 2012.

Simple and modern, the new building's design features an intersection of geometric shapes. Reasonably arranged in its layout, the Archive is divided into three main areas, namely the Archival Repositories, Public Services Area and Office Area, which are effectively linked and independent in functions. They are composed of Archival Repositories, Multi-purpose Lecture Halls, Exhibition Halls, Conference Rooms, Showrooms, Digital Processing Center, Fire Monitoring Center, Computer Management Center, Information Reference & Use Center, Office Area, etc.

The Archives is equipped with intelligent management systems and modern management facilities as well as equipment like advanced automatic fire extinguishing system, automatic sprinkler system, automatic fire/burglar alarm system, electronic monitoring system, central air conditioning system, fingerprint identification access control system, purification and disinfection machines, humidification and dehumidification machines, etc.

The Archives has become a landmark in the suburbs of Yangquan new city. It is a modern public archives playing multiple roles of being an archives and documents storage base, a patriotism education base, a social practice base for archiving education in primary and secondary schools, a prevailing documents access and use center, a government information center, electronic document center and government conference center, reflecting the concept of public archives construction.

2 | 3
4 | 5
6

2 库房 / Repository
3 档案查阅室 / Archives Reference Room
4 展厅 / Exhibition Hall
5 中控室 / Central Control Room
6 多功能报告厅 / Multi-purpose Lecture Hall

垣曲县档案馆

Yuanqu Archives

一层平面图 / Ground Floor Plan

占地面积： 3400m²
建筑面积： 3560m²
库房面积： 1260m²
层　　数： 地上 5 层，地下 1 层
设计单位： 运城市博博建筑设计有限公司
建成时间： 2016 年 5 月

Site area: 3,400m²
Floor area: 3,560m²
Repository area: 1,260m²
Number of floors: ↑ 5 floors and ↓ 1 level
Time of completion: May 2016

垣曲县档案馆成立于1980年，截至2018年底，馆藏各门类档案86610卷（件）。

垣曲县档案馆新馆位于垣曲县东环路，2016年5月建成投入使用，所处环境优美，设计新颖，设施完备，功能齐全。

新馆建筑在外形上充分体现档案元素，立面虚实结合、形式优美；内部采用中间走廊、南北两侧布置房间的长方形布局方式，简洁清晰、分区明确。新馆由档案库房、档案技术用房、对外服务用房以及办公用房四部分组成。公共服务区域包括位于一层的档案查阅利用大厅、开放档案电子档案阅览室、政府公开信息查阅中心、展览大厅以及位于六层的学术报告厅；二层是办公区；档案库房分置于地下一层、三层和四层；档案技术用房包括档案接收、整理编目、抢救修复、消毒除尘、数字化加工用房等，位于三层北侧及五层。

档案库房装备了600m³的密集档案柜、防磁柜、底图柜，配备了库房门禁、防盗、防火、温湿度自动控制系统和档案数字化管理系统，能够满足今后30年20万卷馆藏发展的需要。实现了档案安全保管基地、爱国主义教育基地、档案利用中心、政府信息利用查阅中心、电子文件利用中心功能的完美结合，为垣曲档案事业的发展奠定了坚实的基础。

1 档案馆外景 / The Archives Exterior

Yuanqu Archives was established in 1980. As of the end of 2018, the Archives has a collection of 86,610 files (items) of different classifications.

The new site of Yuanqu County Archive is located at Donghuan Road, Yuanqu County. Completed and put into use in May 2016, it enjoys beautiful environment, novel design, superb facilities and complete functions.

The new Archives' contour is designed to fully reflect archive-related elements, with a facade that embodies both virtuality and reality in a graceful appearance. Inside the building, a rectangular layout is adopted with corridors in the middle and rooms at the south and north wings, presenting a simple while clear structure and explicitly-separated areas, namely the Archival Repositories, Technical Rooms, Public Services Area and Office Area. The public services area consists of the Archives Reference and Use Room, Public Electronic Archive Reference Room, Government Information Reference Center, and Exhibition Hall on 1F, as well as the Academic Lecture Hall on 6F. On 2F is the Office Area. Archival Repositories are on B1, 3F and 4F respectively. Located on the north side on 3F and 5F, Technical Rooms are divided by different purposes including archive receiving, sorting and cataloging, rescue and repair, disinfection and dust removal, digital processing, etc.

The Archival Repositories are equipped with compact shelving, anti-magnetic shelving and base drawing shelving that occupy $600m^3$ in total, furnished with Repository access control, anti-theft, fire prevention, auto temperature/humidity control systems and archive digital management system, guaranteeing a capability to preserve 200,000 files for the next 30 years. The Archives is a perfect integration of being an archives storage base, a patriotism education base, an archives access and use center, a government information center, and an electronic document center. It has laid a solid foundation for the development of the archive sector of Yuanqu County.

2
3 | 4
5

2 多功能厅 / Multi-purpose Hall
3 展厅 / Exhibition Hall
4 库房 / Repository
5 一层大厅 / GF Lobby

鄂尔多斯市东胜区档案馆

Dongsheng District Archives, Erdos

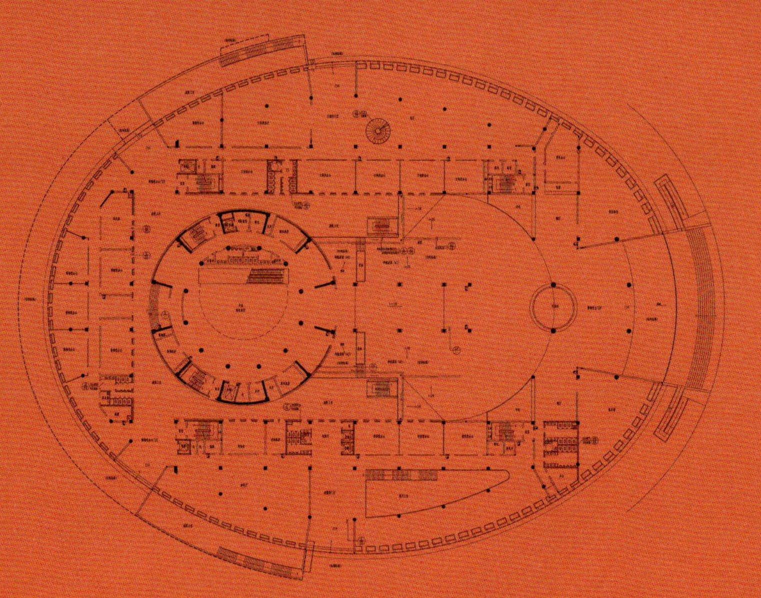

一层平面图 / Ground Floor Plan

占地面积： 55915m^2
建筑面积： 5973m^2
库房面积： 2620m^2
层　　数： 地上 2 层，地下 1 层
设计单位： 中国建筑西北设计研究院有限公司
建成时间： 2014 年 11 月

Site area: 55,915m^2
Floor area: 5,973m^2
Repository area: 2,620m^2
Number of floors: ↑ 2 floors and ↓ 1 level
Time of completion: November 2014

旧馆 Old

New 新馆

鄂尔多斯市东胜区档案馆始建于1952年，1999年11月晋升为内蒙古自治区一级档案馆，2010年晋升为国家二级档案馆。现存文书、会计、婚姻、公证、诉讼等各门类档案共334714卷（件），其中文书档案124450卷（件），资料22741册，数据库录入案卷级目录77718条，文书档案数字化率已达到58%。

鄂尔多斯市东胜区档案馆新馆为合建项目，与东胜区博物馆和文保所集中于一栋建筑中。三部分相互独立，分别设置对外出入口，内部又通过展厅、咖啡厅等开放空间紧密联系。整个建筑地下1层，地上6层，其中档案馆部分位于地下1层、地上2层。档案馆位于南侧，由一层入口进入后，右侧是通高的开放展览大空间，宽敞明亮；左侧是与博物馆共用的咖啡厅。地下布置有档案库房，一层、二层北侧分别为档案技术与办公、会议用房。

新馆建筑由厚重敦实的裙房、层层相错的主楼、轻盈通透的穹顶三个部分组成。主楼错动的环形削弱其体量感的同时，丰富了室内外的空间层次，给人以盘旋上升、灵动飘逸的观感。主楼中间通高的中厅，上盖半球状的玻璃穹顶，是全国最大的弧形玻璃天窗。穹顶上镶嵌的卷云纹金边，是从“匈奴王冠”中获得灵感，经现代建筑手法重新解构，呈现出具有独特雕塑感的“王者之冠”建筑形象。整体建筑既传统又现代，既高贵典雅又质朴大气，现已成为东胜区地标性建筑，在蓝天白云与绿地的映衬下熠熠生辉。

Dongsheng District Archives was founded in 1952. In November 1999, it was upgraded to a class-1 Archives of Inner Mongolia Autonomous Region, and in 2010, a national class-2 Archives. The Archives has a total of 334,714 files (items) in all classifications, like administrative, accounting, marriage, notarization, litigation, etc. Among them there are 124,450 files (items) of administrative archives and 22,741 volumes of documentary archives. In its database there are 77,718 file-level category records, and 58% of the administrative archives have been digitalized.

The new building of Dongsheng District Archives is a joint construction, which also accommodates Dongsheng District Museum and Relics Conservation Institute in the same building. The three organizations are independent of each other with separate entrances available for each of them, while inside they are interconnected closely through open spaces like exhibition hall, cafeteria, etc. The whole building has 1 basement level and 6 floors aboveground, among which the Archives has one basement level and 2 floors aboveground. The Archives is on the south side of the building. After we enter the Archives through the entrance on 1F, an open space for exhibition in full-height can be seen on the right side, spacious and well-lighted. On the left, there is a coffee shop between it and the museum. The Archives Repository is placed underground, and on the north side of 1F and 2F are Technical Rooms and Office/Conference Rooms respectively.

The new building consists of three parts: the thick and solid podium building, the main building with floors on split levels, and the light while transparent dome. The split-leveled ring of the main building weakens its sense of volume and also enriches the spatial levels of indoor and outdoor structures, offering a sense of circling and moving. The full-height hallway in the middle of the building is covered with a hemispherical glass dome, the largest curved glass skylight in China. The gold rim of cirrus clouds on the dome is inspired by the Xiongnu Crown, the treasure of the Archives. Restructured in modern architectural approach, it presents an architectural image "Crown of the King" with a unique sculptural conception. The whole building is traditional while modern, noble, simple and elegant. It has become a landmark building of Dongsheng District, shining against the blue sky, white clouds and green grass.

1 档案馆外景 / The Archives Exterior

2
3 | 4
5 | 6

2 建筑外立面局部 / Partial Building Façade
3 一层大厅 / GF Lobby
4 建筑外观 / Building Appearance
5 库房 / Repository
6 档案查阅室 / Archives Reference Room

新巴尔虎右旗档案馆

New Barag Right Banner Archives

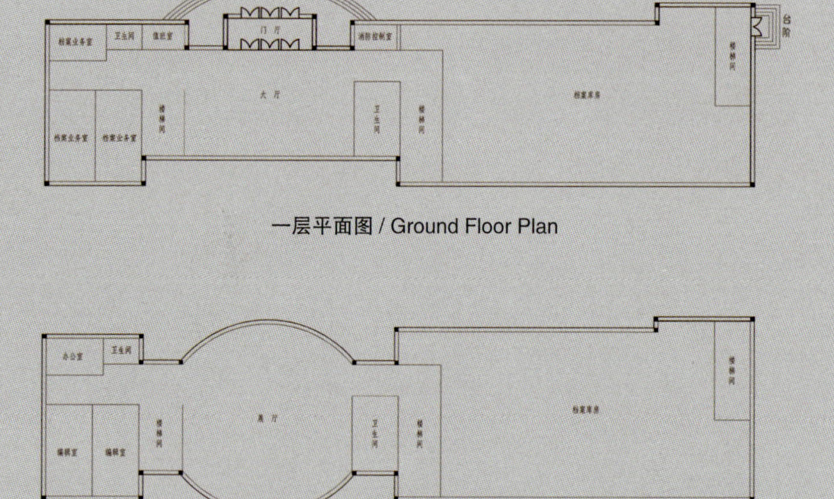

一层平面图 / Ground Floor Plan

二层平面图 / Second Floor Plan

占地面积： $4200m^2$
建筑面积： $3007m^2$
库房面积： $980m^2$
层　　数： 4 层
设计单位： 呼伦贝尔勘查设计院
建成时间： 2015 年 1 月

Site area: $4,200m^2$
Floor area: $3,007m^2$
Repository area: $980m^2$
Number of floors: 4 floors
Time of completion: January 2015

旧馆 Old

New 新馆

新巴尔虎右旗档案馆馆藏档案总量达到 21558 卷又 49462 件（共计 71020 卷件），另有声像档案 2148 张、255 盘，实物档案 192 个，资料 1499 本，共 73 个全宗、6 个门类。

新巴尔虎右旗档案馆新馆为框架结构的独立建筑，布局紧凑，主要由馆库区、多功能区、业务办公区三个部分组成。馆库区分为声像实物档案库、文书档案库、备用库，其中文书档案库 3 个，库房总面积 980m²；多功能区包括展厅、多功能会议室、接待大厅，其中会议室与展厅各 130m²；业务办公区主要集中在第三层。

新馆建筑立面采用两种颜色的外挂石材，分别与玻璃窗相结合，层次分明。入口大厅、展厅与会议大空间为镶嵌在立方体建筑中间的圆柱形体，圆柱体上空设置钢架穹顶构筑物，强调入口主要空间的同时勾勒出优美的建筑轮廓。

The total collection of New Barag Right Banner Archives have reached 21,558 files and 49,462 items (71,020 files/ items in total), 2,148 audio/ video records, 255 disks, 192 material objects, and 1,499 documentary archives, totaling 73 archive fonds in 6 classifications.

The new Archives is a separate building with a framework structure and compact layout. It mainly consists of three parts, namely the Repository Area, Multi-functional Area and Business & Office Area. The repository area is divided into audio/video repositories, administrative repositories, and standby repositories. The total area of administrative archive repositories are 980m². The multi-functional area consists of the Exhibition Hall, Multi-purpose Conference Room, and Reception Hall. Among them, the conference room and exhibition hall occupies 130m² respectively. The business & office area is mainly on 3F.

The facade of the new building is made of stones in two colors on the exterior side. They integrate with glass windows separately to show distinct layers. Large spaces like the entrance hallway, Exhibition Hall and conference space are cylindrical spaces embedded in the middle of the cubic building. A steel frame dome structure is used above the cylinder, which offers elegant outlines of the building while emphasizing the main space of the entrance.

1 档案馆外景 / The Archives Exterior

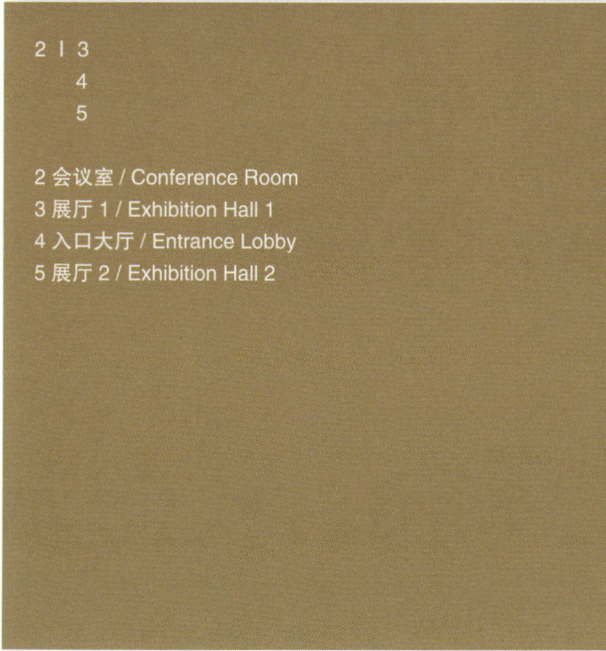

2 | 3
4
5

2 会议室 / Conference Room
3 展厅 1 / Exhibition Hall 1
4 入口大厅 / Entrance Lobby
5 展厅 2 / Exhibition Hall 2

沈阳市于洪区档案馆

Yuhong District Archives, Shenyang

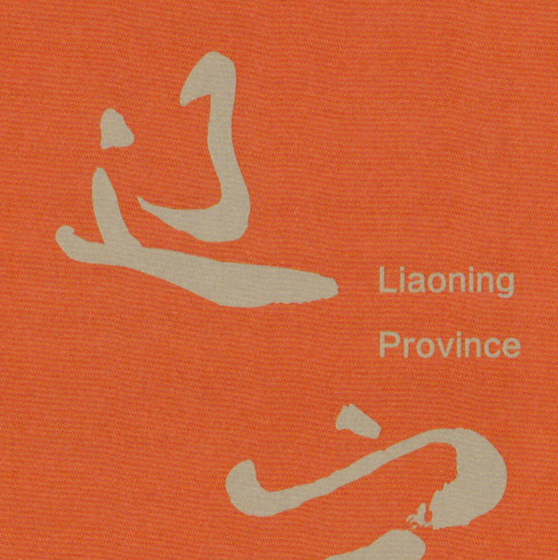

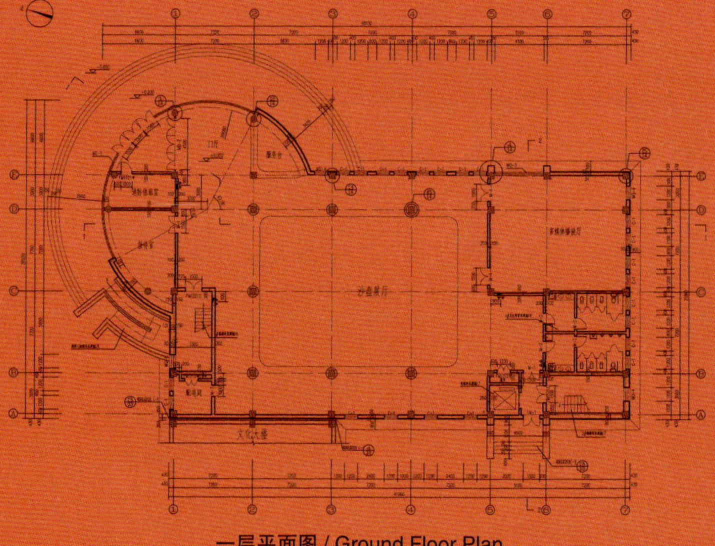

一层平面图 / Ground Floor Plan

占地面积： 4235m^2
建筑面积： 5221m^2
库房面积： 1500m^2
层　　数： 主体 5 层
设计单位： 沈阳新都工程咨询有限公司
建成时间： 2012 年 3 月

Site area: 4,235m^2
Floor area: 5,221m^2
Repository area: 1,500m^2
Number of floors: 5 floors for the main building
Time of completion: March 2012

沈阳市于洪区档案馆始建于1979年7月，现有馆藏143个全宗，各种门类和载体档案17大类11余万卷，2012年晋升为国家一级档案馆。

档案馆新馆为独立建筑，外部与区文化大楼连接，地上共5层，总建筑面积5221m^2，包括现代化库房、档案查阅大厅、报告厅、展厅、数字加工中心、电子文件中心及办公等功能区。其中展厅、档案查阅大厅等对外开放区位于一层、二层，方便群众参观展览及查阅利用，沙盘展厅位于建筑中心的一、二层通高空间，开敞明亮，三、四、五层主要为封闭的档案库房、功能用房以及办公区域，保障了馆藏档案安全及日常工作的独立运行。档案馆整体功能分区明确、流线清晰。

新馆建筑立面采用实体与玻璃幕墙相结合的方式，重点突出，浑然一体，成为和谐广场旁一座标志性建筑。

Yuhong District Archives of Shenyang was initially built in July 1979. Now it has 143 fonds, and more than 110,000 files of archives of 17 classifications in various categories and carriers. In 2012, it was upgraded to a national class-1 archives.

The new Archives is an independent building, connected to the district cultural building from outside. It consists of five floors with a total floor area of 5,221m^2, including modern repositories, archives reference halls, lecture hall, exhibition hall, digital processing centers, electronic document center and office areas. Among the rooms, areas that are open to the public like the exhibition hall and archives reference hall are on 1F and 2F so the public can conveniently visit the Exhibition Hall for referencing and utilizing the archives. Open and bright, the sand table exhibition hall is a full-height space that occupies 1F and 2F; 3F, 4F and 5F are mainly closed archive repositories, functional rooms and office areas, which secure the holding and uninterrupted daily operation. Overall functions of the Archives are well-defined with clear streamlines.

The facade of the new building is a combination of physical masses and glass curtain walls, with features highlighted in integral building. It has already become a landmark next to Hexie Square.

1 档案馆外景 / The Archives Exterior

沁县档案馆
Qin County Archives

一层平面图 / Ground Floor Plan

占地面积： 3266m²
建筑面积： 3052m²
库房面积： 1067m²
层　　数： 主楼 3 层
设计单位： 山西省第二建筑设计院
建成时间： 2017 年 11 月

Site area: 3,266m²
Floor area: 3,052m²
Repository area: 1,067m²
Number of floors: 3 floors for the main building
Time of completion: November 2017

沁县档案馆成立于 1958 年 11 月，馆藏档案资料 15 万卷（件、册），包括文书、业务、会计、科技、声像、实物六大类。

新馆位于县城南湖公园西南角，是一座设计新颖独特、馆藏结构合理、设备先进、门类齐全、功能完善的现代化国家综合档案馆。

新馆建筑简洁大方，采用 U 形空间布局。一层正中设置入口服务大厅，东西两翼分别为实物展厅与影像展厅，对沁县的历史与现状进行展示。为进一步强化档案馆的公共服务功能，馆内还开设了档案查阅室、公开阅览室、现行文件查阅室、多媒体报告厅等馆室，分设于 U 形空间二、三层中间连通处，方便公众到达，东西两翼为馆库。

馆库分为智能化、实物馆库、密集架、档案备份 4 个库区，全部安装了温湿度自动控制系统、消防自动灭火系统、视频监控系统，档案保管水平全面提升。

沁县档案馆突出了“面向未来、面向大众”的新理念，兼具时代感和地方特色，以现代建筑语言与装饰风格来解读这座城市的百年记忆，彰显出北方水城特有的厚重历史积淀和独特文化韵味。沁县档案馆成为国有档案资源和社会档案资源兼备的档案资源库和服务百姓、服务社会的重要平台，为全市档案馆建设积累了宝贵的经验，为全省的县级新馆建设起到了示范作用。

1 档案馆外景 / The Archives Exterior

珲春市档案馆

Hunchun Archives

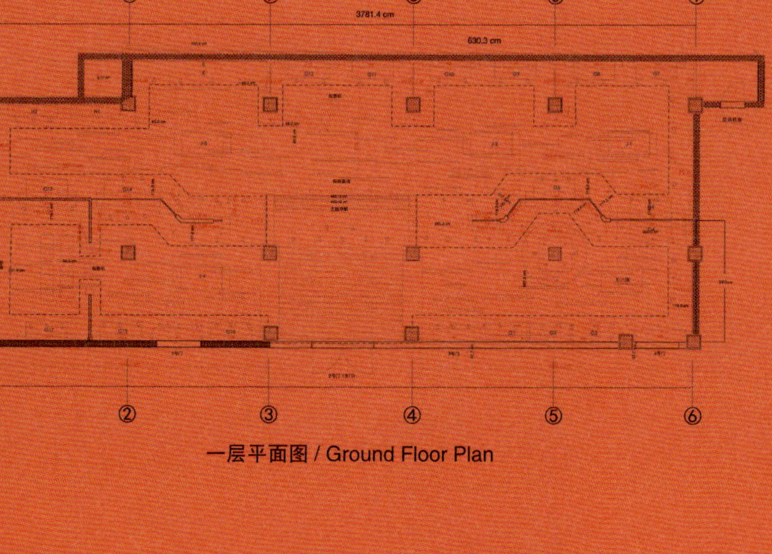

一层平面图 / Ground Floor Plan

占地面积：6000m^2
建筑面积：5786m^2
库房面积：1984m^2
层　　数：6 层
设计单位：北京清尚环艺建筑设计院有限公司
建成时间：2016 年 11 月

Site area: 6,000m^2
Floor area: 5,786m^2
Repository area: 1,984m^2
Number of floors: 6 floors
Time of completion: November 2016

珲春市档案馆主要负责收集和保管市直党政机关、团体、乡镇、企事业等单位形成的具有永久、长期保管价值的各类档案资料。珲春市档案馆新馆于 2016 年 11 月建成投入使用。

新馆位于珲春市新城区站前大街，毗邻珲春高铁站、珲春市法院、珲春市公安局，与市规划馆、图书馆、青少年活动中心合建形成建筑组群，统称“三馆一中心”，作为珲春市城市文化中心。其中档案馆为主楼，居于正中。

档案馆建筑地上共分 6 个区域，不同区域分层管理。一层为地情资料馆、档案修复室、机房等技术用房区，二层为档案管理科、查阅利用区及公共服务区，三层为档案展览区，四层为档案库房区和档案整理、消毒、接收区，五层为档案库房区，六层为办公服务区。分区布局合理，各区域利用效率高。

“三馆一中心”建筑群采用“U”形组合方式，形成半围合的入口广场。各馆形体自成一体又紧密联系，统一而不失变化。建筑风格宏伟大气，浑然一体，成为珲春市地标式文化建筑，承载着这座城市的历史、当下与未来。

1 档案馆外景 / The Archives Exterior

Hunchun Archives mainly collects and preserves all kinds of archives with permanent and long-term values from the municipal party and government organs, groups, townships, enterprises, institutions, etc. The new archives was completed and put into use in November 2016.

Located at Zhanqian Street of Hunchun's Xincheng District, the new building is adjacent to Hunchun High-speed Railway Station, Hunchun Municipal Court, and Hunchun Municipal Public Security Bureau. It is part of the building cluster which also comprises the Municipal Planning Museum, Library and Youth Activity Center, together with them referred to as "Three Halls and One Center" as the urban cultural center of the City. Among the buildings, the Archives is the major one in the center.

The building is divided into six areas aboveground, with different areas arranged and managed on different floors. On 1F are the technical rooms including Geographic Information Room, Archive Restoration Room, Computer Rooms, etc. On 2F are the Archive Management Department, Reference and Use Area and Public Service Area. On 3F is the Archives Exhibition Area. On 4F are the Archive Repository Area and Archive Sorting, Disinfection, and Accession Areas. On 5F is the Archives Repository Area, and on 6F is the Office and Service Area. The areas are in reasonable partitioned and efficiently utilized.

The "Three Halls and One Center" building complex is in U-shape to form a semi-enclosed entrance to the square. The halls are both self-contained and closely connected, unified while diversified. In grand and harmonious architectural style, it has become the landmark cultural building of the city, carrying the history, present and future of Hunchun City.

2
3 | 4
5 | 6

2 展厅 / Exhibition Hall
3 接待室 / Reception Room
4 展廊 / Exhibition Gallery
5 会议室 / Conference Room
6 档案查阅室 / Archives Reference Room

2
3 | 4
5 | 6

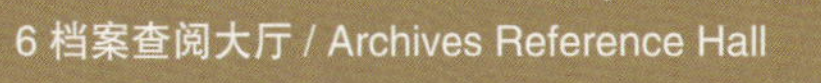

2 展厅 1 / Exhibition Hall 1
3 报告厅 / Lecture Hall
4 展厅 2 / Exhibition Hall 2
5 二层展廊 / 2F Exhibition Gallery
6 档案查阅大厅 / Archives Reference Hall

龙井市档案馆

Longjing Archives

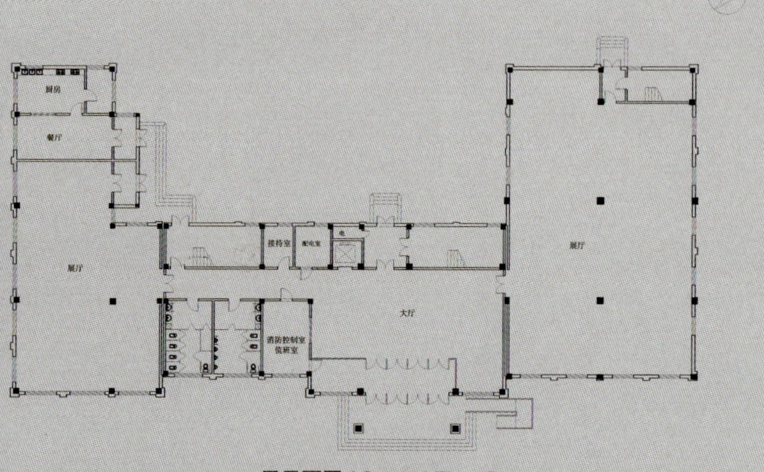

一层平面图 / Ground Floor Plan

占地面积： 9098m²
建筑面积： 4680m²
库房面积： 1733m²
层　　数： 主楼 4 层
设计单位： 延边东北亚建筑设计院有限公司
建成时间： 2018 年 12 月

Site area: 9,098m²
Floor area: 4,680m²
Repository area: 1,733m²
Number of floors: 4 floors for the main building
Time of completion: December 2018

龙井市档案馆成立于 1959 年 7 月 5 日，2012 年被评为国家二级档案馆，馆内两个展厅先后被评为市、州爱国主义教育基地、廉政教育基地、民族团结进步教育基地、少数民族文化传承基地。

龙井市档案馆新馆位于延龙路以西、吉林职业技术学院东北，2017 年 4 月开工建设，2018 年 12 月建成，占地面积 9098m^2，建筑面积 4680m^2。

新馆建筑采用 U 形布局，地上 4 层。U 形空间中间连通处西侧为服务空间，布置有楼梯、卫生间，被服务空间紧紧围绕着服务空间。一、二层为对外开放的入口大厅、展厅、接待室、餐厅、查阅中心及文化体验馆；三层是库房、档案业务与技术用房；四层为库房及办公用房、报告厅。馆内建有《韩乐然生平展》，展陈面积共计 400m^2，展线 141m。

档案馆新馆作为龙井市的文化展示利用窗口，其建筑结合当地民俗风格融入朝鲜族大屋顶元素，外立面开窗整齐规律，外观端庄典雅，融古贯今。

1 档案馆外景 / The Archives Exterior

Longjing Archives was established on July 5, 1959. In 2012, it was awarded "National Class-2 Archives". Two exhibition halls in the Archives have been elected patriotism education bases of the city and prefecture, Yanbian Prefecture Integrity Education Base, National Unity and Progress Education Base, and Ethnic Minority Culture Inheritance Base.

The new building is located in the west of Yanlong Road and the northeast of Jilin Vocational and Technical College. Its construction was started in April 2017 and completed in December 2018. It covers 9,098m^2 of land and the building area is 4,680 square meters.

The new building takes a U-shaped layout with four stories aboveground. The west side and the connected middle part of the U-shaped space are service space, where there are stairs, toilets, and service space that is surrounded by other service space. 1F and 2F are open to the public: entrance hall, exhibition hall, reception room, restaurant, reference center and cultural experience hall; on 3F are repositories, archives business and technical rooms; on 4F are repositories, offices and lecture hall. There is "Han Leran Life Exhibition" in the museum. The exhibition takes 400m^2 and the exhibition line is 141 meters.

As a cultural window of the city, the new building puts local folk style into Korean roof elements. Neat and regular windows are arranged on the facade. Its appearance is dignified, elegant, traditional, yet modern.

2
3 | 4
5 | 6

2 展厅 1 / Exhibition Hall 1
3 档案文化体验中心 / Culture Experience Center
4 大厅 / Archives Hall
5 档案查阅接待大厅 / Archives Reference/Reception Hall
6 展厅 2 / Exhibition Hall 2

通化县档案馆

Tonghua Archives

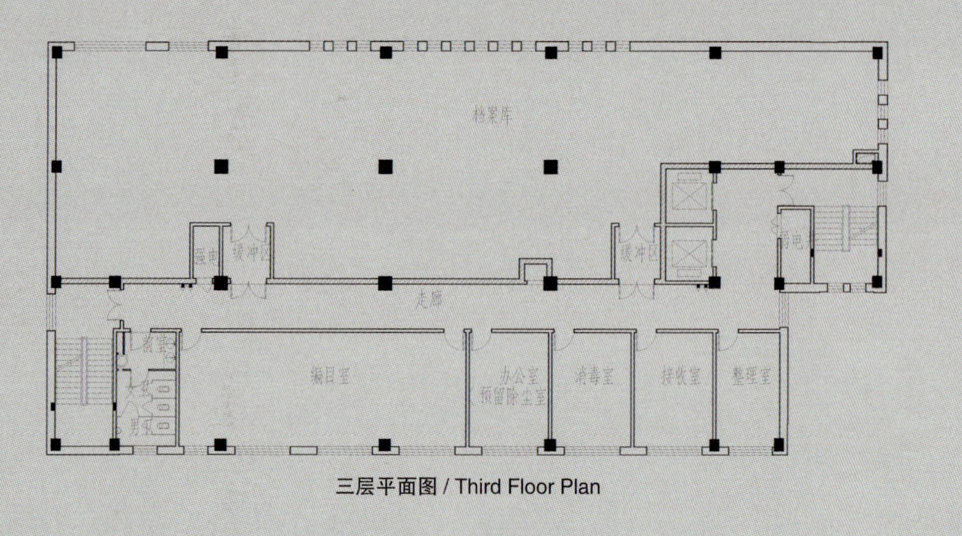

三层平面图 / Third Floor Plan

占地面积：1569m²
建筑面积：5087m²
库房面积：2687m²
层　　数：6 层
设计单位：吉林省建苑设计集团有限公司
建成时间：2018 年 11 月

Site area: 1, 569 m²
Floor area: 5,087m²
Repository area: 2,687m²
Number of floors: 6 floors
Time of completion: November 2018

通化县档案馆成立于 1963 年，馆内珍藏自 1877 年设县以来两千多卷清代和近万卷民国时期的县衙文书档案，同时还保有新中国成立后十类档案共 26 万多卷。

通化县档案馆新馆于 2016 年动工，2018 年投入使用。新馆坐落于快大茂镇长征路 862 号，地上 6 层，主要由档案保管、展示利用与办公三部分组成。其中对外开放的服务大厅、展厅与查阅中心位于一、二层，方便公众到达；库房与档案业务技术用房位于三、四、五层；六层为办公、会议空间，整个建筑布局合理、流线清晰。

新馆建筑在形式处理上，与室内功能相结合。一层挑出的入口雨棚、外凸的墙体和楼板均起到了对入口及开放空间的提示作用；三 ~ 五层库房部分采用实墙与竖向条形玻璃；二层大面积玻璃的使用，使得视觉上成为一层的延续而同时与三层脱离；顶层办公区域则为大面积采光的横向窗户。

通化县档案馆新馆集爱国主义教育基地、档案安全保管基地、档案利用中心、政府公开信息查询中心、电子文件备份中心等功能为一体，有力地推动了地方档案事业的发展。

1 档案馆外景 / The Archives Exterior

Tonghua County Archives was established in 1963. The Archives has collected more than 2,000 files of administrative archives from the Qing Dynasty since it became a county in 1877 and nearly 10,000 of files of administrative archives from the Republic of China. There are also over 260,000 files in 10 classifications after the founding of the People's Republic of China.

Construction of the new building was started in 2016 and it was put into use since 2018. The new building is located at No.862, Changzheng Road, Kuaidamao Town. It has six stories aboveground and is mainly composed of three parts, namely archive custody, exhibition, access and use, and offices. The Service Hall, Exhibition Hall and Reference Center that are open to the public are on 1F and 2F, where can be conveniently accessed by the public; repositories, archival business and technical rooms are located on 3F, 4F and 5F; 6F is the office and conference space. The entire building's layout is reasonable and the streamlines clear.

The new building's design combines the external appearance with indoor functions. The entrance canopy extended out of the building, the convex wall and the floor slabs on 1F remind the public of the entrance and public space; the repositories from 3F to 5F adopt solid walls and vertical strip glasses; the use of large-area glasses on 2F offers visual continuation to 1F and at the same time separate the floor from 3F; in the office area on the top floor there are large-area horizontal windows for better lighting.

The New building of Tonghua Archives provides multiple functions including patriotism education base, secure archives preservation base, archive access and use center, government public information reference center, and electronic file backup center, and has effectively promoted the development of the local archives sector.

2 | 3
4
5 | 6

2 档案查阅接待大厅 / Archives Reference/Reception Hall
3 中控室 / Central Control Room
4 报告厅 / Lecture Hall
5 档案查阅室 / Archives Reference Room
6 库房 / Repository

甘南县档案馆

Gannan Archives

Heilongjiang Province

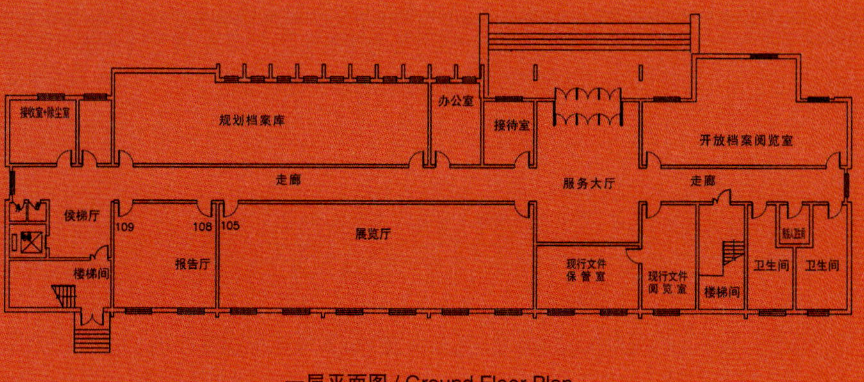

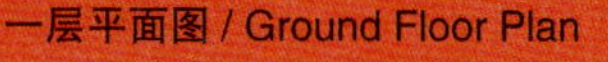
一层平面图 / Ground Floor Plan

占地面积：2500m²
建筑面积：3642m²
库房面积：2105m²
层　　数：主楼 4 层、配楼 1 层
设计单位：齐齐哈尔市垦业建筑设计有限公司
建成时间：2016 年 10 月

Site area: 2,500m²
Floor area: 3,642m²
Repository area: 2,105m²
Number of floors: 4 floors for the main building, and 1 floor for the wing
Time of completion: October 2016

甘南县档案馆于1958年9月成立，至今已有50多年的馆龄。截至2018年，馆内共保存县属各单位档案98个全宗55000卷、6000余册图书资料。馆藏档案资料记载了甘南县的历史变迁，包括文书、科技、专门、声像、实物等多门类、多载体档案。

甘南县档案馆新馆于2015年9月开始建设，2016年10月竣工，是一栋4层的框架结构建筑，由档案库房、业务技术用房与对外服务区三个主要部分组成，功能齐全、布局合理。一层对外服务用房除展厅外，还设置有兼具档案查阅登记、政府信息公开、电子文件、音像档案阅览、复印室等功能于一体的阅览室；二层为业务和技术用房，包括档案接收、除尘、消毒室，档案整理、编目、档案数字化用房等；三、四层档案库区包含有文书档案库、专业档案库、音像档案库、特殊载体档案库和图书资料室等。

新馆建筑形象与功能有机整合，以诠释档案馆的功能性、艺术性、文化性和时代性，从形体表皮纹理、色彩、材质及尺度控制上折射出现代建筑中的文化意韵。主体墙面由数字“50”围合而成，适度表达出档案馆的历史年代与文化性。西侧消防泵房错落有致、浑然一体，婉约地体现了现代建筑韵味。整组建筑简洁大气，庄重典雅，不仅具有强烈的时代气息，还概括地体现了建馆历史与文化，尽显当代档案馆的信息殿堂之韵。

1 档案馆外景 / The Archives Exterior

Gannan Archives has been serving for more than 50 years since its establishment in September 1958. As of 2018, a total of 98 fonds including 55,000 files from the county's various organizations and more than 6,000 books have been kept in the archives. The holdings record the historical changes of Gannan County, including different classifications of archives on all types of carriers such as administrative archives, science and technology archives, specialized archives, audio and visual archives, and material objects.

With its construction being commenced in September 2015 and completed in October 2016, the new building is a 4-storey framework structure consisting of three main parts, namely Archive Repositories, Business and Technical Rooms and Public Service Area. It is well functioned with a reasonable layout. 1F accommodates public service rooms including the Exhibition Hall, and reading room that offers multiple functions including archive referencing and registration, government information disclosure, electronic documents, audiovisual archive review, and photocopying. On 2F are business and technical rooms for archives accession, dust removal, disinfection room, archive sorting, cataloging, archive digital rooms, etc. 3F and 4F are archives repositories including administrative archives repository, specialized archives repository, audio/visual archives repository, special carrier archives repository and library.

The image and functions of the new building are properly integrated to interpret its functional, artistic, cultural and contemporary natures. Cultural meaning of the modern architecture is reflected from its texture, color, materials and control of the scale. The walls of the main building are enclosed in the form of number "50", which moderately shows the age and cultural importance of the Archives. The fire pump rooms on the west side are randomly scattered while integrated into one, implying the charms of a modern architecture. The entire building complex is simple, elegant, solemn and elegant. It not only displays a strong sense of the time, but also embodies the history and culture of the Archives, showing the rhythm of the contemporary archives as a hall of information.

2
3 | 4
5

2 一层大厅 / GF Lobby
3 会议室 / Conference Room
4 库房 / Repository
5 资料室 / Documentary Archives Room

桦南县档案馆

Huanan Archives

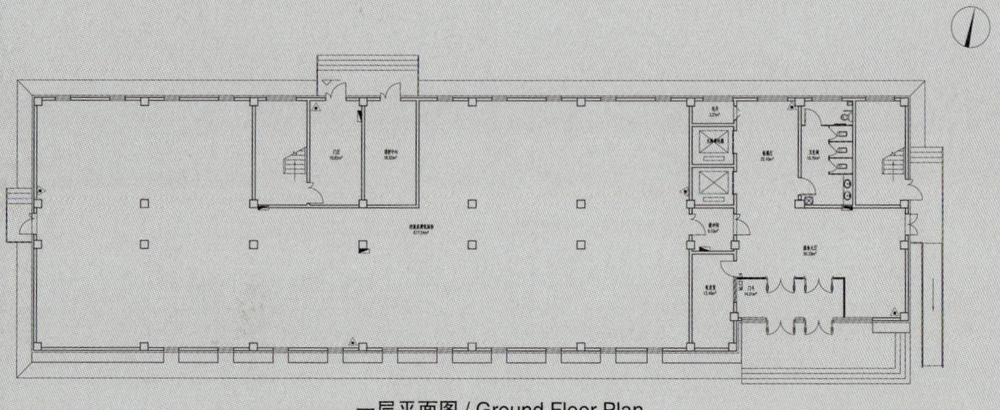

一层平面图 / Ground Floor Plan

占地面积：2272m²
建筑面积：4019m²
库房面积：1434m²
层　　数：5 层
设计单位：东北林业大学工程咨询设计研究院有限公司
建成时间：2017 年 11 月

Site area: 2,272m²
Floor area: 4,019m²
Repository area: 1,434m²
Number of floors: 5 floors
Time of completion: November 2017

桦南县档案馆馆藏档案 104 个全宗，63427 卷，各种资料 15370 册，图片资料 3600 张，影像资料 128 本，公章 16 枚，奖杯奖状 25 件，家谱 10 本。

桦南县档案馆新馆位于苗圃路与育才街交汇处，于 2016 年正式动工，2017 年 11 月竣工并投入使用，是目前黑龙江省内规模最大的县级现代化档案馆之一。

新馆建筑采用长方形布局，地上 5 层。主入口位于长方形长边的端侧，保证了主立面及内部空间的完整性。一层、二层为档案库房，有单独的档案入口及档案专用电梯，库房总面积约 1434m^2；对外开放服务区位于三层，包括餐厅、接待室、阅览室、展厅、报告厅等，设有自入口直达的电梯与楼梯；档案业务与技术用房位于四层；五层为办公用房。

建筑外观为方整厚重的几何体，入口所在的一角处理为内凹的玻璃幕墙，立面正常的开窗之外置有一层档案密码样式的符号，凸显着档案馆的文化气息。

桦南县档案馆集档案保管基地、现行文件查阅中心、档案编研利用中心、电子文件管理中心、爱国主义教育基地为一体，有效满足了人民群众查阅人事、婚姻等民生档案及编史修志、工作查考对档案的利用需要，在服务经济建设及社会发展等方面发挥积极作用。

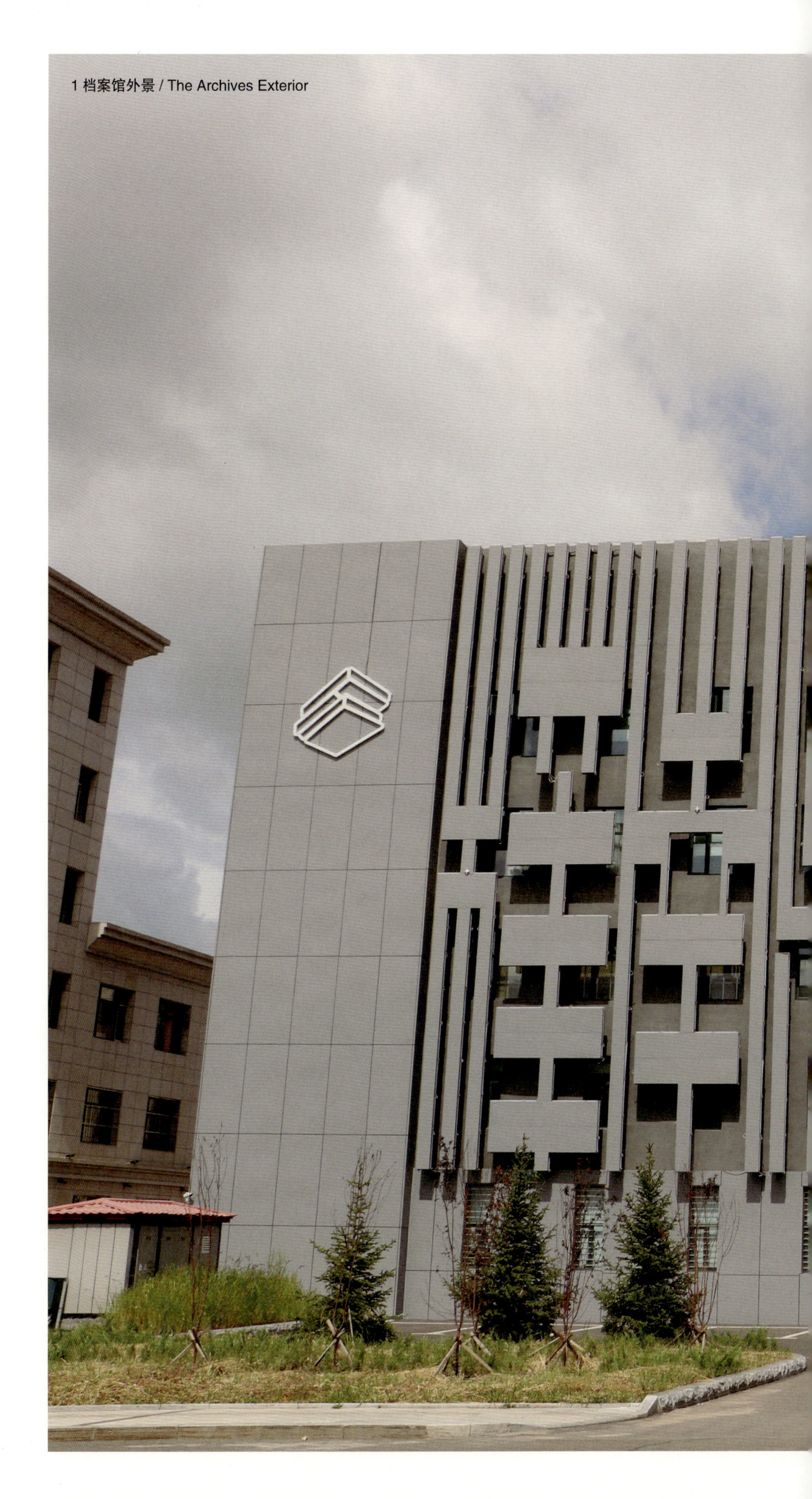

1 档案馆外景 / The Archives Exterior

Huanan Archives has a collection of 104 fonds including 63,427 files, 15,370 volumes of various documentary archives, 3,600 photos, 128 books of videos, 16 official seals, 25 cups and awards, and 10 genealogies.

The new building of Huanan Archives is located at the intersection of Miaopu Road and Yucai Street. The construction was officially started in 2016 and completed and put into use in November 2017. Currently, it is one of the largest county-level modern archives in Heilongjiang Province.

The new building is in a rectangular layout with five floors aboveground. The main entrance is at one end of the long side, ensuring the integrity of the main facade and interior spaces. 1F and 2F accommodate Archives Repositories, with separate archive entrances and dedicated elevators are available. The total area of the repositories is about 1,434m^2. The Public Service Area is on 3F, including the Restaurant, Reception Room, Reading Room, Exhibition Hall, Lecture Hall, etc. Elevators and stairs that are directly accessible from the entrance are available. Archives Business and Technical Rooms are on 4F, and 5F is the office space.

The building takes the appearance of a bulky geometry. At the corner where the entrance is located is a concave glass curtain wall. Where there are windows on the facade, there are a layer of symbols that look like archive passwords, which highlight the cultural atmosphere of the Archives.

Playing various roles as an archives custody base, prevailing document reference center, archives compilation, research and utilization center, electronic document management center, and patriotism education base, Huanan Archives effectively meets the demands of the public to find reference to archives related to people's well-being like personnel matters, marriage, etc., as well as the needs for archives in historical research and work review. It has exerted a positive role in serving the economic construction and social development.

2
3 I 4
5

2 一层大厅 / GF Lobby
3 展厅 / Exhibition Hall
4 库房 / Repository
5 档案查阅室 / Archives Reference Room

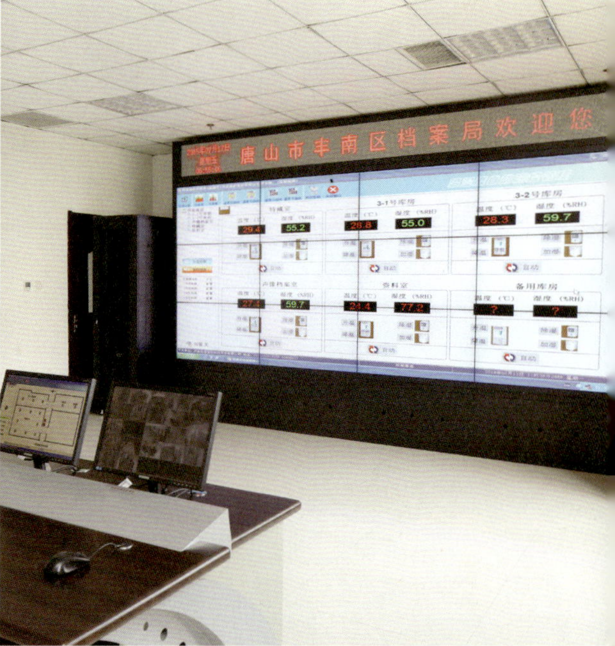

2 | 3
4
5 | 6 | 7

2 一层大厅 / GF Lobby
3 中控室 / Central Control Room
4 档案查阅大厅 / Archives Reference Hall
5 声像室 / Audio/Video Room
6 库房 / Repository
7 资料室 / Documentary Archives Room

豐南區

河北
Hebei
Province

旧馆
Old
New
新馆

霍山县档案馆

Huoshan Archives

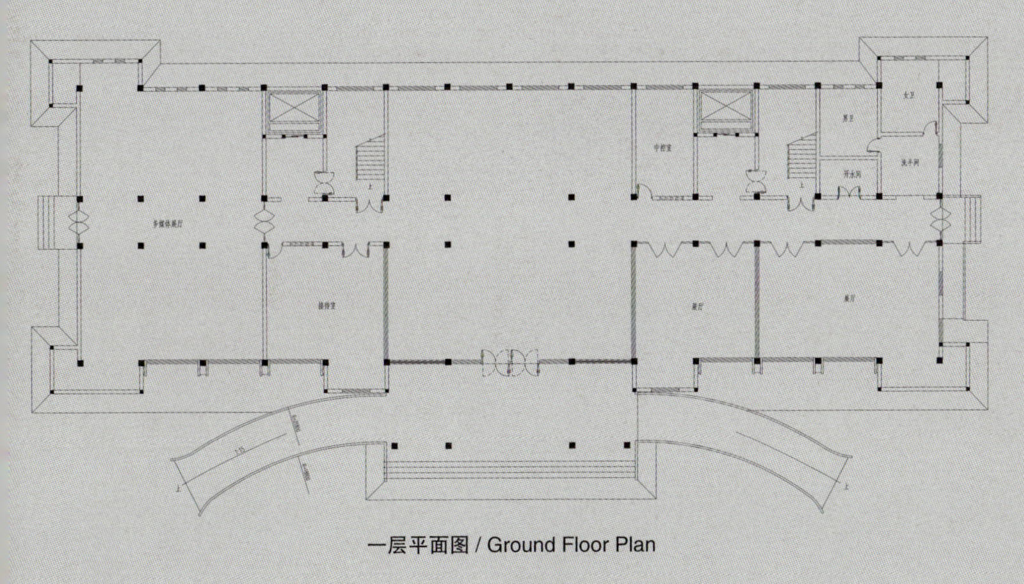

一层平面图 / Ground Floor Plan

旧馆 Old

New 新馆

占地面积：4467m²
建筑面积：4699m²
库房面积：1853m²
层　　数：5 层
设计单位：霍山天地建筑勘察设计院
建成时间：2011 年 12 月

Site area: 4,467m²
Floor area: 4,699m²
Repository area: 1,853m²
Number of floors: 5 floors
Time of completion: December 2011

霍山县档案馆成立于1959年，2014年1月晋升为国家二级综合档案馆，同年被评为省级中小学生档案教育实践基地和市级爱国主义教育基地。

霍山县档案新馆于2011年12月建成并投入使用，为全框架独栋建筑，主体5层，严格遵循档案库房的设计原则。馆内功能齐全，设有中央控制室、档案库房（包括特藏室）、展厅、档案利用服务中心（包括电子文件查阅区、政府公开信息查阅区、开放档案查阅利用区等）、计算机中心、业务技术用房等。其中展厅、档案利用服务中心位于一层和二层，方便群众使用；三～五层布置档案库房、档案业务与技术用房、办公用房。

馆内设施一流，温湿度控制系统、自动防火防盗报警系统、视频监控系统、门禁系统、消毒灭菌系统等一应俱全；配备了中央空调、公用电梯和档案专用电梯、智能密集架、防磁柜、服务器、存储器等设施设备，充分体现了现代化档案馆的多功能性、开放性和服务性。

新馆建筑墙体自下而上内收，吸收古代城墙做法，屋顶采用中国传统坡屋顶样式，立面为典雅的三段式，开窗比例优美，呈现出汉代风格，体现着档案馆作为文化建筑的历史厚重感。

1 档案馆外景 / The Archives Exterior

Huoshan Archives was established in 1959. In January 2014 it was upgraded to a national class-2 archives, and in the same year was elected a provincial archives education practice base for primary and secondary schools and municipal patriotism education base.

The new building of Huoshan Archives was completed and put into use in December 2011. It is a full-frame single building with five floors in the main part, designed in strict accordance with the principles for archives repositories. The archives provides all-round functions by providing the Central Control Room, Archive Repositories (including Special Collection Room), Exhibition Hall, Archives Access and Use Service Center (including Electronic Document Reference Area, Government Public Information Reference Area, Public Archives Reference and Utilization Area, etc.), Computer Center, Business and Technological Rooms, etc. The Exhibition Hall and Archives Utilization Service Center are on 1F and 2F to facilitate public access; archives repositories, archives business, technical rooms, office rooms are on 3F to 5F.

The building is equipped with first-class facilities, including temperature and humidity control system, automatic fire/burglar alarm system, CCTV system, access control system, disinfection and sterilization system, etc. Facilities and equipments like central air conditioners, public elevators and dedicated archives elevators, compact shelving, anti-magnetic cabinets, storage, etc. are available, fully reflecting the versatility, openness and services of a modern archives building.

Walls of the new building are concaved from bottom to top to follow suit of ancient city walls, and the traditional Chinese pitched roof style is adopted. The facade is in a three-part elegant style, with artistic window opening ratio showing the style of the Han Dynasty. All these reflect the historical importance of an archives as a cultural building.

2 | 3
4
5

2 一层大厅 / GF Lobby
3 展厅 / Exhibition Hall
4 移动展厅 / Mobile Exhibition Hall
5 档案利用服务大厅 / Archival Access and Use Hall

泗县档案馆

Sixian Archives

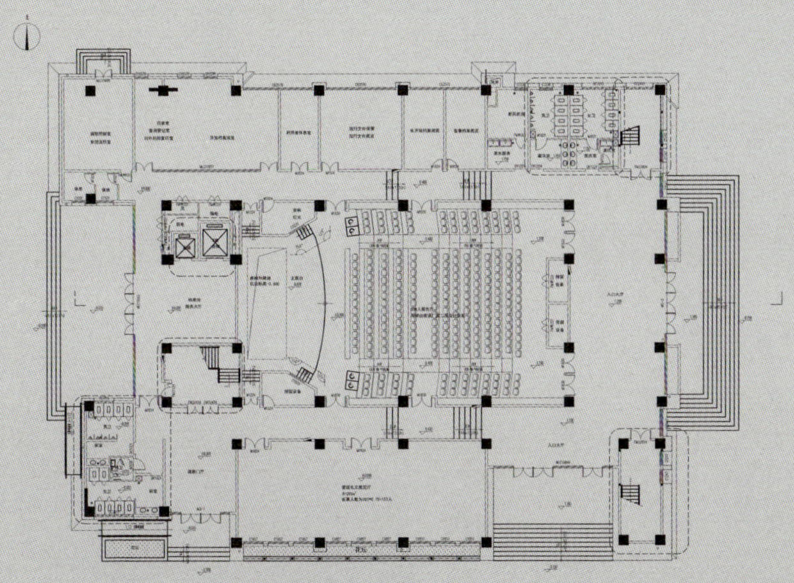

一层平面图 / Ground Floor Plan

占地面积： 10000m²
建筑面积： 7298m²
库房面积： 1884m²
层　　数： 主楼 4 层
设计单位： 安徽省城建设计研究总院有限公司
建成时间： 2018 年 7 月

Site area: 10,000m²
Floor area: 7,298m²
Repository area: 1,884m²
Number of floors: 4 floors for the main building
Time of completion: July 2018

泗县档案馆新馆坐落于泗县人民政府的西北角，收集了泗县自新中国成立后到现在的所有珍贵档案和史料。建筑为框剪结构，地上 4 层，呈“回”字形布局：“回”字中心一～三层为报告厅与库房大空间，四层为三层屋顶，其他房间围绕“回”字周边布置。一层设有入口大厅、档案服务大厅、展厅、报告厅及档案登记阅览室等对外开放功能，方便群众到达使用；二层为会议与办公用房；三层集中布置档案库房，包括智能库房、普通档案库房、特藏库房；四层则为档案业务、技术用房。

新馆设计前泗县档案工作人员与设计单位密切接洽多次，使得设计充分考虑了泗县地方的风土人情特色，立面紧扣庄重、典雅、现代的设计理念，主楼通过横向体块的分割弱化原来的比例，大面积石材的应用体现其文化性和历史厚重感。整个建筑气势恢宏匠心独具，既透着古朴又充满朝气。

泗县档案馆新馆的建设有力地推进了地方档案事业的发展。

1 档案馆外景 / The Archives Exterior

Located in the northwest corner of the People's Government of Sixian County, Sixian Archives has a collection of all the precious archives and historical documents of the county since the founding of People's Republic of China. The building adopts a frame-shear structure with 4 floors aboveground. Its layout is like the Chinese character "回": In the center of the building, large spaces of the Lecture Hall and Repositories on 1F to 3F constitute the inner square of the character; 4F is the roof of 3F, and the other rooms are arranged around the center. On 1F there are facilities opened to the public like the Entrance Hall, Archive Service Room, Exhibition Room, Lecture Room and Archive Registration and Reading Room, which are easy to access; on 2F are the conference and office rooms; 3F is dedicated for archives repositories, including the Intelligent Repository, Ordinary Archives Repository and Special Collection Repository; 4F accommodates Archive Business and Technical Rooms.

Before the new building was designed, the staff had been in close contact with the design team for many times so that the local customs and characteristics of the county can be taken into full consideration. As a result, the facade fits the solemn, elegant and modern design concept, and the chunky main building, divided by the horizontal masses. The use of stones in large areas reflects the cultural and historical sense of the Archives. The building is magnificent, unique, simple and energetic as a whole.

Construction of the new Sixian Archives has vigorously promoted the development of the local archive sector.

2 | 3
4 |
5

2 会议室 / Conference Room
3 报告厅 / Lecture Hall
4 展厅 / Exhibition Hall
5 库房 / Repository

望江县档案馆

Wangjiang Archives

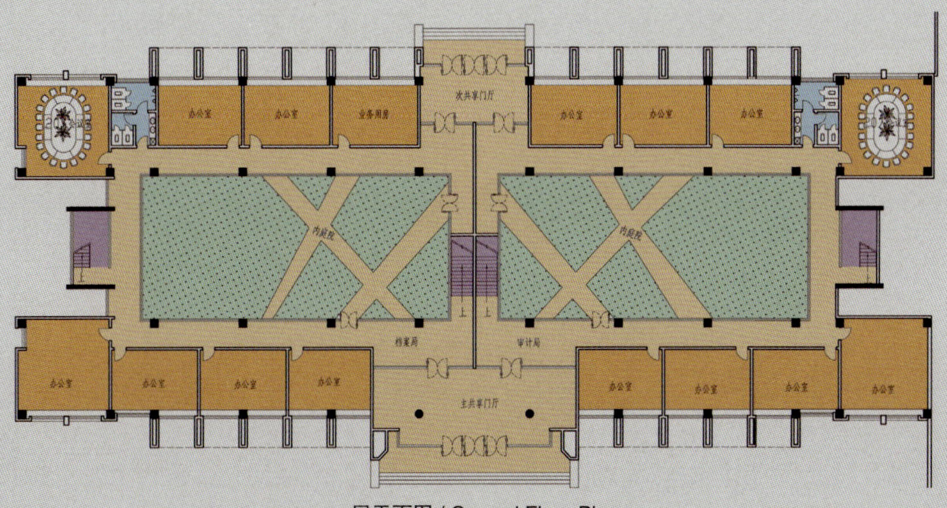
一层平面图 / Ground Floor Plan

占地面积： 14674m²
建筑面积： 5592m²
库房面积： 2174m²
层　　数： 地上 3 层，地下 1 层
设计单位： 南京城理人城市规划设计有限公司、望江县建筑设计院
建成时间： 2013 年 11 月

Site area: 14,674m²
Floor area: 5,592m²
Repository area: 2,174m²
Number of floors: 3 floors and 1 level
Time of completion: November 2013

望江县档案馆成立于1958年11月，望江县档案馆新馆位于望江县县城华阳镇政务新区E区，与审计局合建，2008年6月开始动工建设，2013年11月竣工，总建筑面积5592m^2，地上3层。

建筑平面呈“日”字形布局，内置两个庭院，围绕庭院布置走廊，走廊另一侧为房间。馆内建有机房、中心监控室、电子文件中心、数字化加工中心和档案利用中心；配备安防、门禁、密集架、恒温恒湿、消防、智能照明等基础设施系统集成。

档案馆建筑与县政务新区行政中心、四大班子办公楼、财政局、土地局建筑共同构成组团，分置于两个条式折形建筑中，并列布置。曲折有度的形体，形成丰富多变的连续空间；规则、有节奏的立面限定出院落、半公共与公共空间；开放的入口空间将基地南北贯穿，呈现出亲切开放的态度。

Wangjiang Archives was established in November 1958. Located in Zone E of the Administrative Affairs New District, Huayang Town of Wangjiang County, the new Wangjiang Archives is built together with the Audit Bureau. Construction began in June 2008 and was completed in November 2013. The total floor area is 5,592m^2 with 3 stories aboveground.

With its planar layout looking like the Chinese character“ 日 ”, it has two courtyards. Surrounding the courtyards corridors are designed, with rooms arranged on the other side the corridors. Inside the building there are Core Computer Room, Central Monitoring Room, Electronic Document Center, Digital Processing Center and Archives Access and Use Center. It is furnished with security, access control, compact shelving, constant temperature and humidity, fire protection, intelligent lighting and other infrastructure systems.

1 档案馆外景 / The Archives Exterior

A cluster is formed by the Archives' building, the Administrative Center of County's Administrative New District, the office building for the four Party and government leading groups, the Finance Bureau and the Land Bureau. They are separately deployed in two strip-shaped buildings and arranged side by side. The tortuous shape forms a rich, varied and continued space. The regular, rhythmic facade defines the courtyards, semi-public and public spaces. The public entrance runs through the base from north to south and shows a cordial and open attitude.

2 | 3
4
5 | 6

2 服务大厅 / Service Hall
3 接待室 / Reception Room
4 展厅 1 / Exhibition Hall 1
5 展厅 2 / Exhibition Hall 2
6 中控室 / Central Control Room

晋江市档案馆

Jinjiang Archives

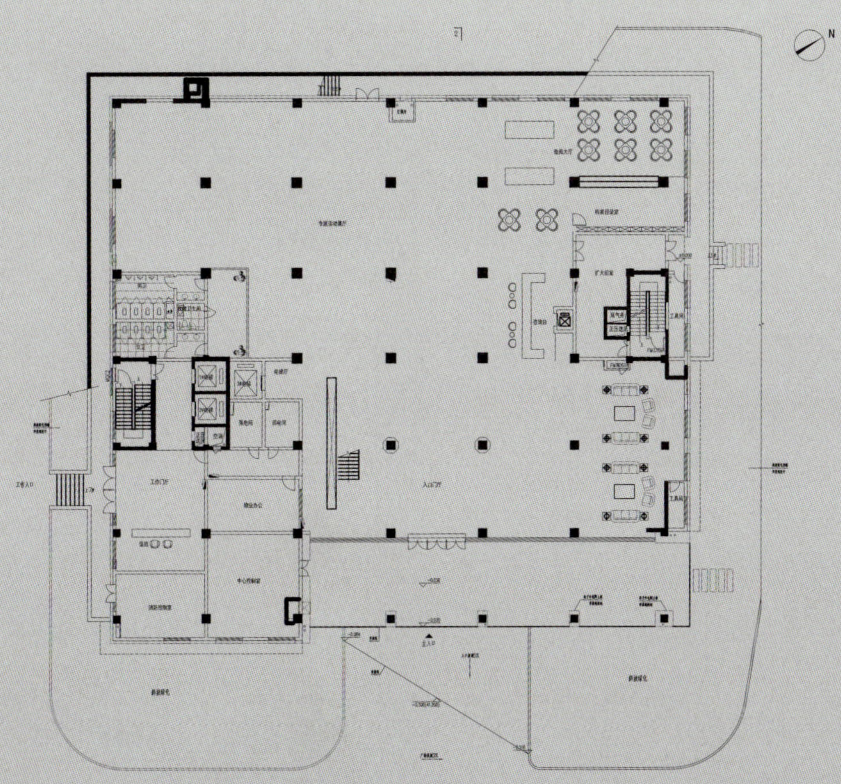

一层平面图 / Ground Floor Plan

占地面积：7402m²
建筑面积：17680m²
库房面积：6000m²
层　　数：地上 9 层，地下 1 层
设计单位：中国水电顾问集团华东勘测设计研究院
建成时间：2013 年 6 月

Site area: 7,402m²
Floor area: 17,680m²
Repository area: 6,000m²
Number of floors: ↑ 9 floors and ↓ 1 level
Time of completion: June 2013

晋江市档案馆成立于 1959 年，档案馆现有馆藏档案 25 万卷，其中文书档案 12 万卷，此外还收有 6000 卷珍贵侨批档案。

晋江市档案馆新馆位于新行政中心核心区，北临十四号路，西临罗裳山，东临人保财险用地，南可俯瞰中心区的市政、市民及文化艺术广场，北可领略优美的罗裳山风景，环境优美。

新馆功能齐全，设置有市民服务区、史实展示区、行政办公区、档案储备区、技术处理区及交流研究区。新馆主入口进入便是两层通高的入口大厅，围绕中庭布置公共服务功能，包括一层的咨询台、休息室、专题活动展厅、档案目录与查阅大厅以及二层的老照片展厅、晋江荣誉展厅。一、二层由门厅左侧的直跑梯连接，呈现出开放流动的建筑空间特性。三～九层中心与北侧均为档案库房，库房有单独入口，并配备两部档案专用电梯。四～七层南侧为档案业务与技术加工用房。办公与会议室位于最高的八层和九层，并设有绿化中庭、绿化庭院，视野开阔，环境舒适，充分体现了设计的人性化。

晋江市档案馆以闽南建筑文化“出砖入石”为脉络，结合地域历史的文化沉淀，充分运用现代空间设计构成手法，合理有序地把各元素进行有机组合。档案馆立面方正现代，外部庄重大气、内部包容开放，是一座集档案科学管理与利用、学术交流、社会教育、政府信息公开及文化休闲为一体的城市标志性公共文化建筑。

1 档案馆外景 / The Archives Exterior

The previous Jinjiang Archives was Established in 1959. The Archives has a collection of 250,000 files among which 120,000 are administrative archives, as well as 6,000 precious Qiaopi files (which are also called Teochew Letters, a unique combination of remittance and correspondence that overseas Chinese sent to their home country in the history).

The new site of Jinjiang Archives is located in the core area of the city's new administrative center, adjacent to No.14 Road on the north, with Luoshang Hill on the west, and PICC's land plot on its east. Overlooking the public, cultural and art square in the center area in the south, and embraced by the graceful scenery of Luoshang Hill in the north, it enjoys pleasant environment.

The new Archives provides complete functions with its six main functional areas, namely the Public Services Area, Heritage Exhibition Area, Administrative Office Area, Archival Repositories, Technical Processing Area and Exchange & Research Area. The main entrance is followed by the well-hole style entrance hall that is two-story high. Public service functions are arranged around the atrium, including the Information Desk, Lounge, Thematic Activities Hall, and the Filing Catalog & Reference Hall on 1F, as well as the Old Photo Exhibition Room and Jinjiang Honors Exhibition Room on 2F. The straight flight staircase at the left side of the hall links F1 with F2, forming an open, flowing architectural space. At the center and north side of 3F to 9F are Archival Repositories, which are accessed through separate entrances and equipped with two elevators for archive use only. On the south side of 4F to 7F are the archives service and technical processing rooms. Offices and conference rooms are on the top two floors, i.e. 8F and 9F, amongst a green atrium and garden courtyard with a humanized design offering the staff with a wide view and comfortable working environment.

Jinjiang Archives was designed by borrowing the "irregularly arranged bricks and stones in proper proportions" from southern Fujian's traditional architectural culture, and giving play to it in combination with the profound local history and culture. The modern spatial design constitutional method was fully utilized, so that a variety of elements were put together in a reasonable, orderly manner in the building. The building's facade is in a modern, upright and foursquare style. Exhibiting a lofty, grand exterior appearance and an inclusive, open atmosphere inside, it is both a public cultural building and city landmark, offering the citizens a venue for archival science management and use, academic exchange, social education, government information disclosure and culture & entertainment.

2 | 3
4 | 5
6

2 党建馆 / Party Building Hall
3 多功能厅 / Multi-purpose Hall
4 展厅 / Exhibition Hall
5 库房 / Repository
6 档案查阅大厅 / Archives Reference Hall

罗源县档案馆

Luoyuan Archives

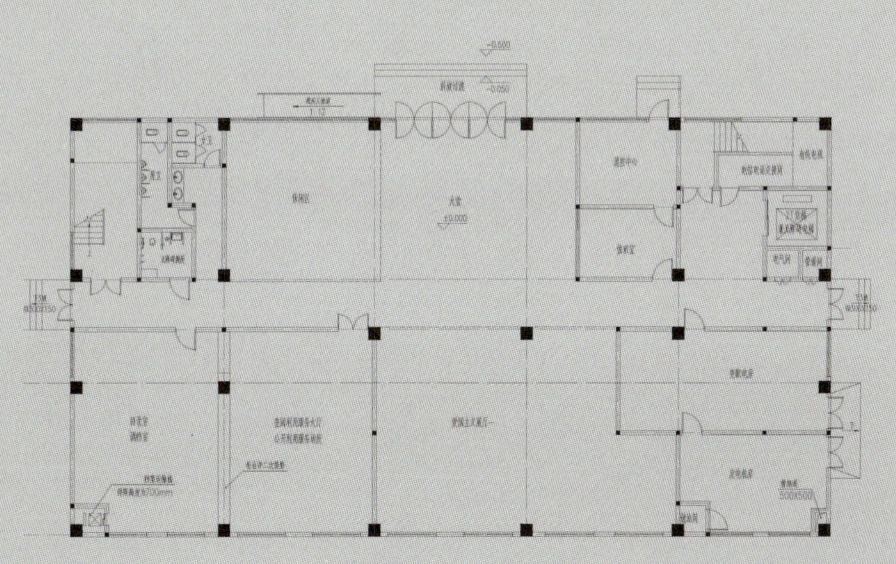

一层平面图 / Ground Floor Plan

占地面积：4203m²
建筑面积：3997m²
库房面积：1326m²
层　　数：主楼 5 层
设计单位：福建省机电建筑设计研究院
建成时间：2017 年 12 月

Site area: 4,203m²
Floor area: 3,997m²
Repository area: 1,326m²
Number of floors: 5 floors for the main building
Time of completion: December 2017

罗源县档案馆于 1958 年 12 月 10 日正式成立，现有馆藏档案 118 个全宗，文书档案 42124 卷、41212 件，民国档案 1380 卷，婚姻档案 42910 件，土地档案 94438 条。

罗源县档案馆新馆位于罗源县江滨南路，2014 年 12 月开工建设，2016 年 3 月 2 日顺利通过竣工验收，2017 年 12 月顺利完成新馆搬迁工作并正式投入使用。

新馆总平面呈长方形，布局简单合理，不同功能用房采用分层布置的方式进行管理。档案查阅、爱国主义展厅等对外服务功能布置在一、二层，方便公众到达；三层为档案业务及技术用房；档案库房集中分布在四、五层，分为密集架纸质档案库房、多媒体档案库房、特藏库等。馆内流线清晰，公众、工作人员与档案接收流线相互独立，避免交叉，保障了档案馆的高效运行。

罗源县档案馆新馆建筑立面现代活泼，独特新颖，在营造现代气息的基础上又注重对历史、文化和开放特征的发掘与表达，通过体块的拉伸、转折等手法营造虚实光影相间的效果及独特文化内涵的建筑形式。

1 档案馆外景 / The Archives Exterior

2 档案馆鸟瞰全景 / The Archives Aerial View

Established on December 10, 1958, Luoyuan Archives has a collection of 118 fonds, 42,124 files and 41,212 items of administrative archives, as well as 1,380 files of holdings from the Republic of China, 42,910 items of marriage holdings, and 94,438 items of land holdings.

The new site of Luoyuan Archives is located on the South Jiangbin Road in Luoyuan County. The construction was commenced in December 2014, and successfully completed and accepted on March 2, 2016. In December 2017, relocation was completed and the new Archives was officially put into use.

The new Archives has a rectangular general layout, which is simple and reasonably designed, with the functional rooms being arranged on different floors for better management. The public service function areas such as archives referencing and patriotism exhibition hall are on 1F and 2F for easy access; archival business and technical rooms are on 3F; and all archival repositories are distributed on 4F and 5F, including compact shelving repositories for paper archives, multimedia archival repositories, special collection rooms, etc. The building has clear, separate streamlines designed for the public, staff members and archives accession, avoiding overlapping and ensuring efficient operation of the Archives.

The building's facade is in a fashionable, lively, unique and novel style, embodying its historical, cultural connotations and openness while creating a sense of a modernism. Through building techniques like stretching and turning of masses, etc., it carries both reality and virtuality, lights and shadows, as well as the unique cultural contents.

3 | 4
5
6 | 7

3 一层大厅 / GF Lobby
4 档案查阅室 / Archives Reference Room
5 多功能会议室 / Multi-purpose Conference Room
6 库房 / Repository
7 资料室 / Documentary Archives Room

上杭县档案馆

Shanghang Archives

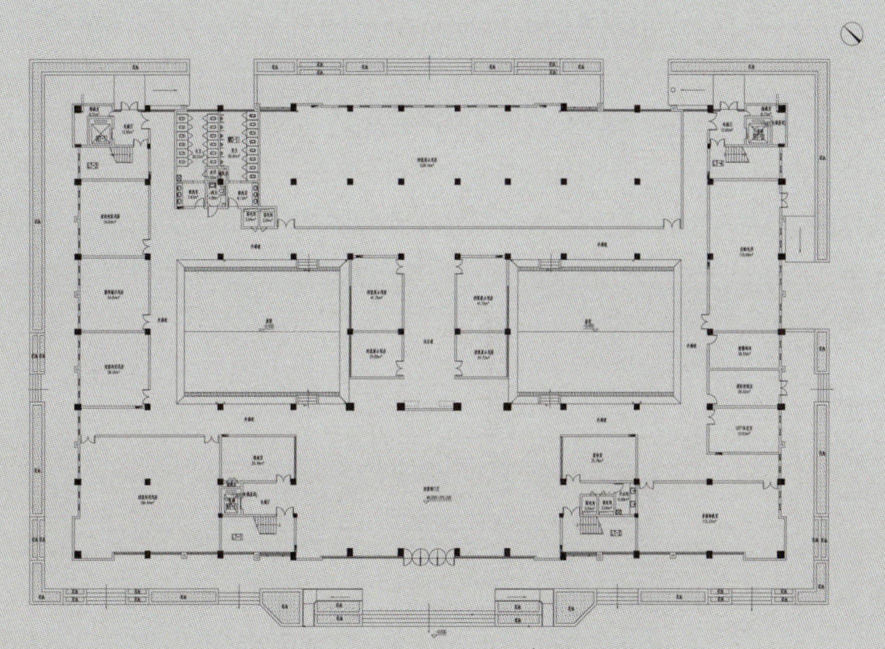

一层平面图 / Ground Floor Plan

占地面积：6500m^2
建筑面积：8938m^2
库房面积：3000m^2
层　　数：主体 3 层
设计单位：中信武汉设计总院
建成时间：2018 年 3 月

Site area: 6,500m^2
Floor area: 8,938m^2
Repository area: 3,000m^2
Number of floors: 3 floors for the main building
Time of completion: March 2018

上杭县档案馆成立于1958年9月3日，现有馆藏文书档案46915卷、21013件，科技档案4848卷，专门档案15706件，资料5699册、各种照片档案3964张，1995年12月，晋升为福建省一级档案馆。

上杭县档案馆新馆位于龙翔新城客家文化中心，在龙翔大道南边石壁寨山脚下，与客家族谱馆、县图书馆、县博物馆、文化馆、科技馆、青少年活动中心等集中建设。

新馆主体建筑共三层，一层以档案接收、消毒、利用、培训、展览功能为主，分布有接收室、消毒室、临时存放室各1间40m^2，多媒体教室1间90m^2，展览空间约1200m^2。二层以档案保管保护功能为主，主要为档案库房共约3000m^2，占总面积的30%，可以接收和容纳未来50年的档案资料，库容量将达到50万卷的规模。三层以档案的整理编目、抢救修复、数字化编研开发和人员办公功能为主，主要为技术用房和办公用房。

整个建筑造型采用围合式布局，形成内置庭院，外立面灰墙、青瓦等民居要素配合玻璃等现代材料，较好地将传统与现代进行结合，既富有客家特色，又体现时代精神，成为上杭县重要的档案资料收集保管基地、档案信息发布利用中心、档案监督管理控制总部、档案文化传播展示平台和档案教育社会实践基地。

1 档案馆外景 / The Archives Exterior

Shanghang Archives was established on September 3, 1958, It has a collection of 46,915 files and 21,013 items, including 4,848 science and technology holdings, 15,706 specialized holdings, 5,699 volumes of documentary archives, and 3,964 photos of all classifications. In December 1995, it was upgraded to a class-1 Archives of Fujian Province.

The new site of Shanghang Archives is located in the Hakka Culture Center of Longxiang New Town, rightly at the foot of Shibizhai Hill to the south of Longxiang Avenue. The construction was carried out as part of the project which also covers the Hakka Pedigree Museum, the County Library, the County Museum, the Cultural Center, the Science and Technology Museum and the Youth Activity Center.

The main building consists of three floors. 1F is mainly for the accession, disinfection, access and use, training, as well as exhibition of archives, comprising of an Accession Room, a Disinfection Room and a Temporary Storage Room, each being $40m^2$, a Multimedia Classroom of $90m^2$, and an exhibition space of about $1200m^2$. 2F is mainly for custody and conservation of archives, with the major part being archives repositories of about $3,000m^2$, accounting for 30% of the total area and being able to acquire and accommodate the archives of the next 50 years, which will bring up the scale of holdings to 500,000 volumes. Functions offered on 3F are chiefly cataloging, rescue and restoration, digital editing, research and development, as well as office affairs, mainly as Technical Rooms and Offices.

The building adopts an enclosed general layout where a built-in courtyard is formed; on the exterior facade traditional residential elements like gray walls and grey tiles are used in perfect combination with modern materials such as glass, reflecting both Hakka characteristics and fashion. The Archives has become an important archival files collection and storage base, archival information release and use center, archival supervision, management and control headquarters, archival culture communication and demonstration platform as well as archives education and social practice base of Shanghang County.

2
3 | 4
5

2 档案馆鸟瞰 / The Archives Aerial View
3 档案查阅大厅 / Archives Reference Hall
4 内院与走廊 / Inner Courtyard and Footpath
5 一层大厅 / GF Lobby

武平县档案馆
Wuping Archives

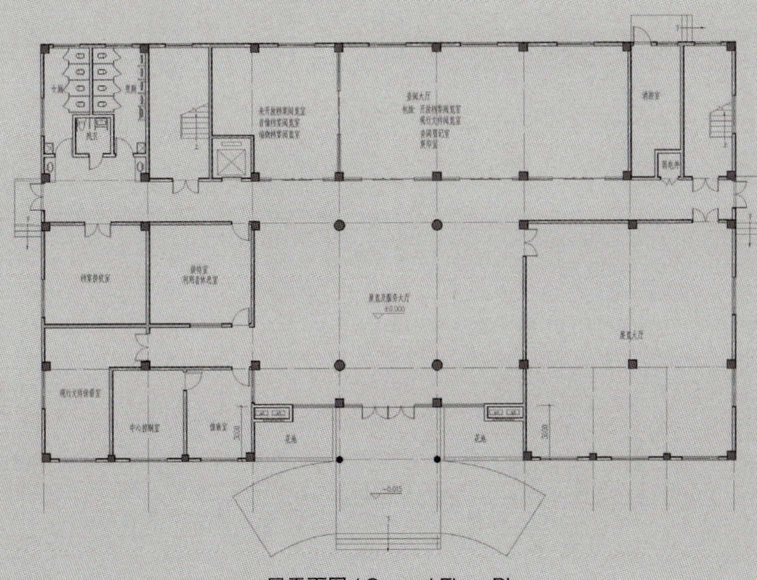
一层平面图 / Ground Floor Plan

占地面积： 3333m²
建筑面积： 4600m²
库房面积： 2643m²
层　　数： 主楼 5 层
设计单位： 福建泷澄集团设计有限公司
建成时间： 2016 年 1 月

Site area: 3,333m²
Floor area: 4,600m²
Repository area: 2,643m²
Number of floors: 5 floors for the main building
Time of completion: January 2016

武平县档案馆成立于1958年10月，现有馆藏档案203个全宗、94929卷、177301件。1994年，武平县档案馆获得省二级档案馆荣誉称号。

武平县档案馆新馆选址于武平县平川镇七坊村北环路段，2013年6月1日正式开工，2015年12月全面完工，2016年1月竣工验收并投入使用。

新馆建筑面积4600m²，共5层，平面采用长方形布局。一层为对外服务与档案接收用房，设有开放的展览及服务大厅、查阅大厅、档案阅览室、接待休息室以及档案接收室、文件保管室，其中对外服务用房面积843m²。二层为办公和业务技术用房，包括办公室、学术报告厅、计算机机房、数字化用房等，其中业务技术用房面积约655m²，办公及辅助用房面积241m²。档案库房分布在三层、四层，五层为预留库房，库房面积共2643m²。

档案馆新馆立面形式典雅，水平的线脚、竖向的柱式以及坡屋顶元素诉说着传统，围绕顶层的蓝色玻璃颇具土楼特征。主入口处内凹处理，虚实对比起到强化突出的作用。整体上，武平县档案馆是一座具有客家元素独特档案文化品位的现代化、公共型、服务型档案馆。

1 档案馆外景 / The Archives Exterior

Established in October 1958, Wuping Archives has 203 fonds, 94,929 files and 177,301 items of archival holdings. In 1994, it was titled Provincial Class-2 Archives.

The new site of Wuping Archives is seated on the North Ring Road section in Qifang Village, Pingchuan Town of Wuping County. The construction was officially started on June 1, 2013, fully completed in December 2015, and accepted for putting to use in January 2016.

The new site has a floor area of 4,600m^2, consisting of 5 floors with a rectangular planar layout. The 1F, designed mainly for public services and archives accession, accommodates open Exhibition and Service Hall, Reference Hall, Archives Reading Room, Reception Lounge, as well as the Archives Accession Room and File Storage Room. Among them public service rooms occupy an area of 843m^2. On 2F there are office spaces as well as business technical rooms, including Offices, Academic Lecture Hall, Computer Room, Digital Room, etc., among which business and technical rooms account for about 655m^2, and offices as well as auxiliary rooms are 241m^2. Archival repositories are distributed on 3F and 4F, while spaces on 5F are reserved for future repositories. The area of all repositories is 2,643m^2 in total.

The building has an elegant facade with horizontal moldings, vertical columns, sloping roofs and other traditional elements. The blue glasses embracing the top floor present characteristics quite similar with those of Fujian Earth Buildings. Large-sized Concave operation are used at the main entrance, highlighting the comparison between reality and virtuality. Wuping Archives on the whole is a modern, public and service-oriented archives with unique Hakka culture elements.

2 | 3
4
5

2 展厅 / Exhibition Hall
3 库房 / Repository
4 阅览室 / Reading Room
5 档案查阅大厅 / Archives Reference Hall

崇义县档案馆

Chongyi Archives

Jiangxi Province

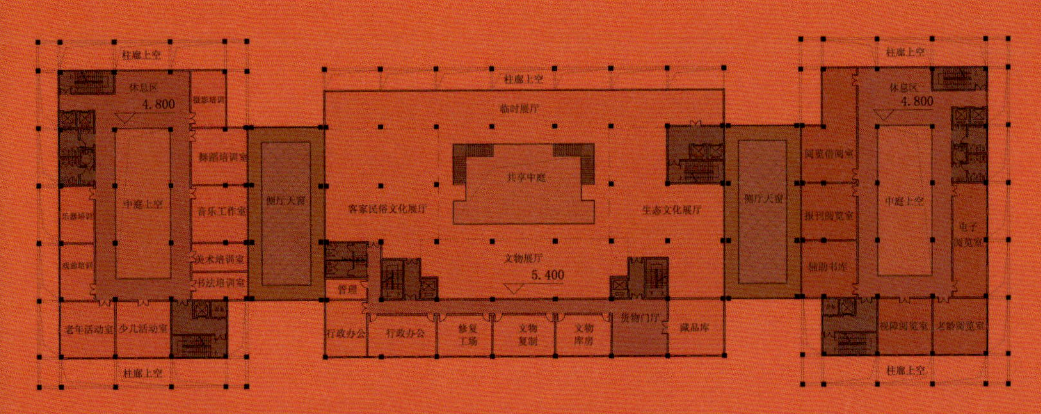

三层平面图 / Third Floor Plan

旧馆 Old

New 新馆

占地面积： 8000m^2
建筑面积： 2839m^2
库房面积： 1500m^2
层　　数： 2 层
设计单位： 杭州中宇建筑设计有限公司
湖南省南方人文化创意有限公司
建成时间： 2018 年 2 月

Site area: 8,000m^2
Floor area: 2,839m^2
Repository area: 1,500m^2
Number of floors: 2 floors
Time of completion: February 2018

崇义县档案馆成立于1959年，现有档案资料51821卷、74262件，其中民国档案8639卷。

崇义县档案馆新馆与县王阳明博物馆、县文化馆、县图书馆、县城市规划展示馆、县妇女儿童活动中心合建，于2015年3月开工建设，2018年2月建成交付使用。新馆建筑呈对称式布局，功能分区明确，排列紧凑，流线清晰。展览部分布置在建筑中心区域，两侧分别设置图书馆、活动室等附属空间。动静分区明确，办公与阅览设有单独的出入口，保证了档案馆的高效运作和管理。此外还设有档案查阅厅、多功能厅、监控室、消毒室、装订室等，配有专用电梯、中央空调、气体灭火系统。室内装饰设计古朴典雅，文化气息浓郁。

建筑造型采用对称的几何造型，主体功能相对突出。公共空间面积足量，一层的开放空间进一步强调了建筑的公共属性，契合当代档案馆的人文与开放精神。

1 档案馆外景 / The Archives Exterior

Established in 1959, Chongyi Archives has 51,821 files and 74,262 items of archives, including 8,639 files of archives from the the Republic of China.

Built as part of the project, which also comprises Wang Yangming Museum, the Cultural Center, Library, Urban Planning Exhibition Hall, and Women and Children's Activity Center of Chongyi County, the new building's construction was started in March 2015 and completed and put into use in February 2018. The building has a symmetrical layout with clearly defined functional areas, compact layout and smooth streamlines. The exhibition room is at the central area of the building. Attached spaces such as library, activity room, etc., are on the wings. Dynamic and static areas are clearly separated. Different entrances are available for offices and reading rooms to ensure efficient operation and management of the archives. In addition, there are Archive Reference Hall, Multi-purpose Hall, Monitoring Room, Disinfection Room, Binding Room, etc. The building is also equipped with dedicated elevators, central air conditioning, and fire extinguishing system. The interior design is simple and elegant, with a strong sense of culture.

The building takes a symmetrical geometric shape, with the main building is relatively highlighted. Sufficient public spaces are offered, with the public spaces on 1F further emphasizes the public attributes of the building to adapt to the spirit of humanity and openness that a contemporary archives should have.

2 | 3
4 | 5

2 档案整理室 / Archives Filing Room
3 库房 1 / Repository 1
4 多功能会议室 / Multi-purpose Conference Room
5 库房 2 / Repository 2

高安市档案馆

Gaoan Archives

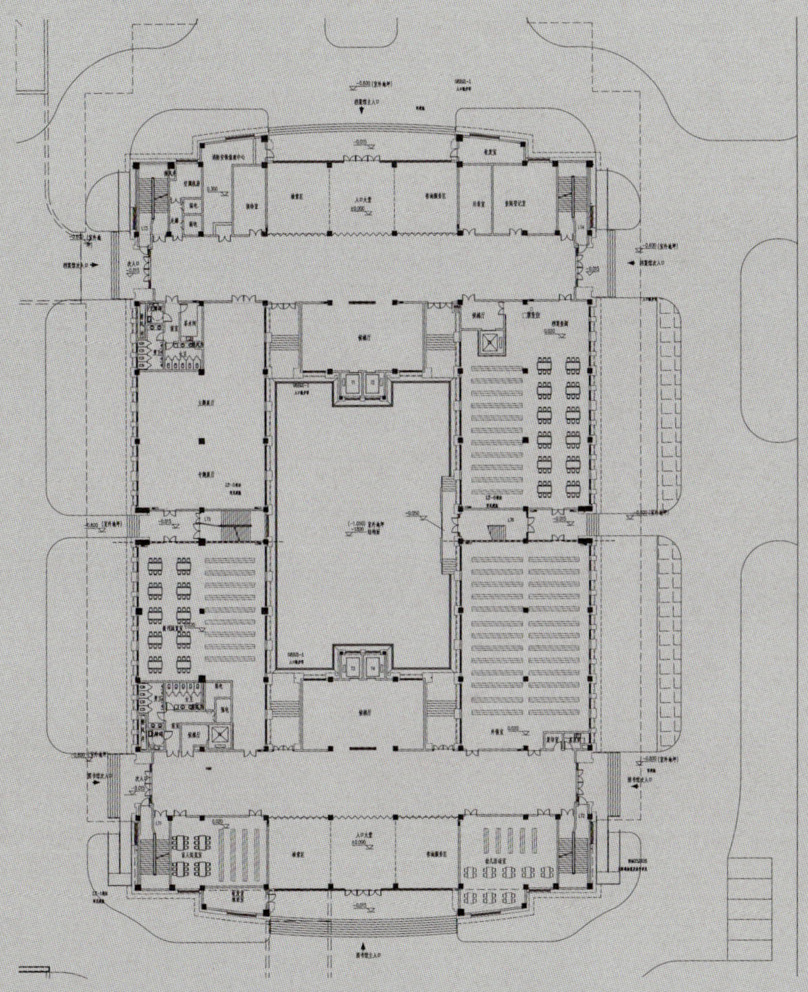

一层平面图 / Ground Floor Plan

旧馆
Old

New
新馆

占地面积：4200m²
建筑面积：12560m²
库房面积：4000m²
层　　数：地上 4 层，地下 1 层
设计单位：上海建工设计研究院有限公司
建成时间：2018 年 1 月

Site area: 4,200m²
Floor area: 12,560m²
Repository area: 4,000m²
Number of floors: 4 floors and 1 level
Time of completion: January 2018

高安市档案馆成立于1959年，馆藏档案资料14.5万卷（册）。

高安市档案馆新馆位于市行政中心西侧，狮子岭公园北侧，与市图书馆合建，2014年3月开工建设，2015年6月主体封顶，2017年3月竣工验收，2018年1月正式投入使用。

新馆排布以中庭大空间为中心，流线设计清晰，功能完善，布局合理。一层为对外服务用房，设有档案查阅大厅、展览大厅和音像演示厅；二层为业务技术用房，设有学术报告厅、数字化加工室、档案整理编目室等场所；三、四层为库房，并设有书画艺术、高安文化名人特藏室。库房的顶层设置保证了档案管理工作相对独立。档案馆设施设备完善，安装了密集架、恒温恒湿系统、高压细水雾灭火系统、红外报警监控系统等。

高安市档案馆建筑造型采用的是简洁的几何形态，立面开窗方式与室内功能相结合，公共空间直贯四层，对建筑的采光、通风、降噪起到了促进作用，一层门厅的设置强调了其建筑开放公共的属性，体现了档案馆的人文与开放性。

高安市档案馆以其具有时代感和现代感的建筑语汇和设计手法，彰显了高安的时代属性，是记录、解读城市的较好诠释。

1 档案馆外景 / The Archives Exterior

Established in 1959, Gaoan Archives has a collection of 145,000 files (volumes).

Located on the west side of the municipal administrative center and north side of Shiziling Park, the new building of Gaoan Archives was built together with the city library. The construction was started in March 2014. The main body was capped in June 2015. The building was completed for acceptance in March 2017, and formally put into use in January 2018.

The new building is centered on the large space of its atrium. With streamlines clearly designed, it has complete functions and a reasonable layout. On 1F, there are rooms for public services, such as the Archives Reference Hall, Exhibition Hall and Audio/Video Demonstration Hall. On 2F are business and technical rooms, including the Academic Lecture Hall; Digital Processing Room, Archives Sorting and Cataloging Room, etc; On 3F and 4F are Archives Repositories, and Special Collection Rooms for painting, calligraphy arts and works of cultural celebrities of Gaoan City. Repositories are arranged on the top floor, maintaining a relatively independent space for the archival work. The Archives is well equipped with compact shelving, constant temperature and humidity system, high-pressure water mist fire extinguishing system, infrared alarm monitoring system, etc.

The building adopts a simple geometric form. The windows on the facade fit the indoor functions. The public space reaches directly to 4F, which enhance the lighting, ventilation and noise reduction. The hallway on 1F enhances the openness, and reflects humanities and openness of the Archives.

With its architectural vocabulary and design techniques full of sense of time and modernity, Gaoan Archives highlights the city's attributes of the times, and is a perfect record and interpretation of the city.

2 | 3
4
5 | 6

2 档案查阅大厅 / Archives Reference Hall
3 多功能厅 / Multi-purpose Hall
4 党员活动室 / Party Member Activity Room
5 一层大厅 / GF Lobby
6 库房 / Repository

井冈山市档案馆

Jinggangshan Archives

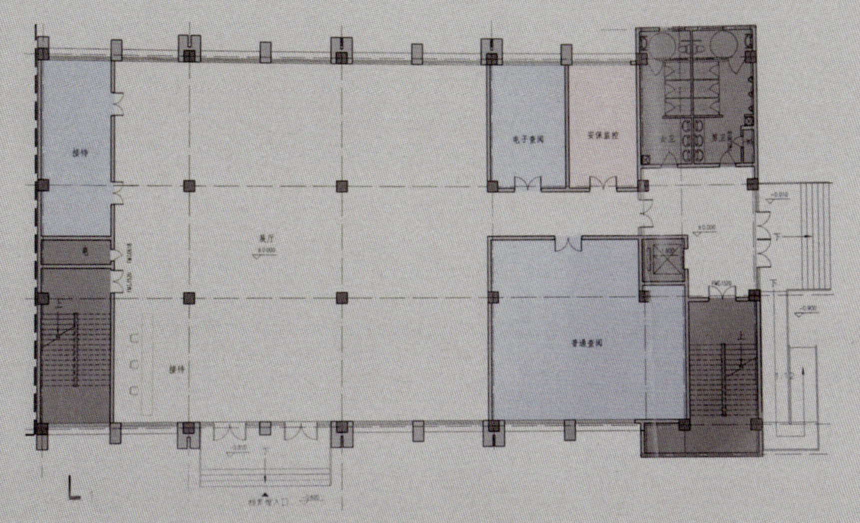

一层平面图 / Ground Floor Plan

占地面积：933m²
建筑面积：3868m²
库房面积：1600m²
层　　数：4 层
设计单位：广东省深圳市建筑设计研究总院
建成时间：2016 年 11 月

Site area: 933m²
Floor area: 3,868m²
Repository area: 1,600m²
Number of floors: 4 floors
Time of completion: November 2016

井冈山市档案馆成立于1959年，目前，馆藏档案资料共5万余卷（册），其中党和国家领导人照片、题词、绘画、手迹等1000余件。

井冈山市档案馆新馆于2011年5月开工建设，2013年完成主体结构，2016年正式投入使用。

新馆建筑功能分区明确，流线清晰，一层为开放大厅，设有档案查阅厅、展览厅及文件查阅等公共服务功能；二层为业务技术用房和办公用房；三层为特藏库、重点档案库房；四层为普通档案库房、资料用房。

档案馆设施设备完善，安装了高压细水雾灭火系统、自动温湿度调控系统、中央空调、档案密集架等。

档案馆建筑形态简洁大方，尽显老区人民艰苦奋斗的作风，立面开窗方式与室内功能一致，统一又富有变化。整体设计手法和谐统一，彰显了老区的文化属性，为井冈山精神的书写描下浓重一笔。

1 档案馆外景 / The Archives Exterior

Jinggangshan Archives was established in 1959. At present, the Archives has a collection of over 50,000 files (volumes), including more than 1,000 items of photographs, inscriptions, paintings and handwritings of Party and national leaders.

The Archives's construction began in May 2011. In 2013, the main structure was completed, and in 2016, it was officially put into use.

Functions of the new building are clearly defined and the streamlines are clearly designed. 1F is a public hall, such as the Archives Reference Room, Exhibition Hall and Documents Reference, etc.; on 2F are business, technology and office rooms; on 3F there are Special Collection Repository and Key Archives Repositories; on 4F there are General Archive Repositories and Documentary Archives Room.

The Archives is well equipped. There are high-pressure water-mist fire extinguishing system, automatic temperature/humidity control system, central air conditioners, compact shelving and so on.

The building takes on a simple while elegant appearance, showing local people's spirit of hard working. The way windows are arranged is consistent with the indoor function, and is both in unity and full of changes. The overall design is harmonious and unified, highlighting the cultural attributes of the city and stressing the Jinggangshan spirit.

2
3 | 4
5

2 展厅 / Exhibition Hall
3 档案阅览室 / Archives Reading Room
4 库房 / Repository
5 档案查阅中心 / Archives Reference Center

万载县档案馆

Wanzai Archives

江西省 Jiangxi Province

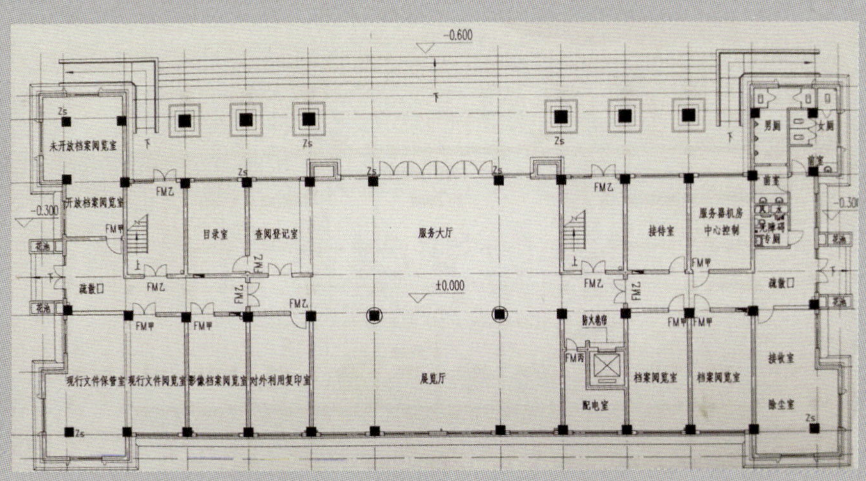

一层平面图 / Ground Floor Plan

占地面积：9986m^2
建筑面积：4700m^2
库房面积：1465m^2
层　　数：5 层
设计单位：浙江东华规划建筑园林设计有限公司
建成时间：2015 年 10 月

Site area: 9,986m^2
Floor area: 4,700m^2
Repository area: 1,465m^2
Number of floors: 5 floors
Time of completion: October 2015

万载县档案馆成立于1959年5月，现有纸质档案14万余卷、8万余件，印章等实物档案2000余件，资料1万余册，2017年12月被江西省档案局授予档案工作规范化管理省一级档案馆称号。

万载县档案馆新馆位于县迎宾大道旁，于2012年10月开工建设，2015年10月建成投入使用。新馆严格按照《档案馆建设标准》和《档案馆建设规范》进行设计施工，坚持“实用、经济、美观”的原则。建筑造型中轴对称，符合档案馆建筑庄重的气质。整体设计采用中国传统建筑园林的创意，既有大气的现代风味，又不乏中国古典韵味，是万载县的标志性建筑之一。

新馆面积充裕、布局合理、功能齐全、设施完备，一层为对外服务用房，二层为档案业务用房和办公用房，三、四层为档案库房，五层为多媒体功能厅。建筑立面统一，一层大厅的通高空间为建筑注入活力，对通风、采光起到了较好的效果，主立面强调建筑的竖向线条与公共属性，满足当代档案馆建筑的人文属性。

1 档案馆外景 / The Archives Exterior

Established in May 1959, Wanzai Archives has more than 140,000 files and more than 80,000 items of paper archives, more than 2,000 material objects such as seals, and more than 10,000 volumes of documentary archives. In December 2017, it was awarded Provincial class-1 Archives by the Jiangxi Provincial Archives Administration for its standardized management of archives.

Located beside Yingbin Avenue in Wanzai County, the new Archives building was started to construct in October 2012, and was completed and put into use in October 2015. The new Archives was designed and constructed in strict accordance with Construction Standards for Archives and Code for Construction of Archives, and following the principle "practical, economical and beautiful". The building adopts an axial symmetrical form to suit its solemnness and grandness. The overall design borrows concepts from traditional Chinese architectures and gardens, which makes it both modern and full of Chinese classical charms. It is one of the landmark buildings of Wanzai County.

The new building has ample floor area, a reasonable layout, complete functions and advanced facilities. On 1F are public service rooms, on 2F are archive business rooms and office rooms, on 3F and 4F are archive repositories, and on 5F is a Multi-media Room. The building has a unified facade, with the full-height space of the halls on 1F filling the building with vitality, as well as sufficient ventilation and lighting. The main facade emphasizes vertical lines of and publicity the building, meeting cultural demands of a contemporary archives building.

2 | 3
4 | 5
6 | 7

2 入口大厅 / Entrance Lobby
3 库房 / Repository
4 展厅 1 / Exhibition Hall 1
5 政府公开信息查阅中心 / Government Public Information Reference Center
6 展厅 2 / Exhibition Hall 2
7 数字化加工用房 / Digital Processing Room

武宁县档案馆

Wuning Archives

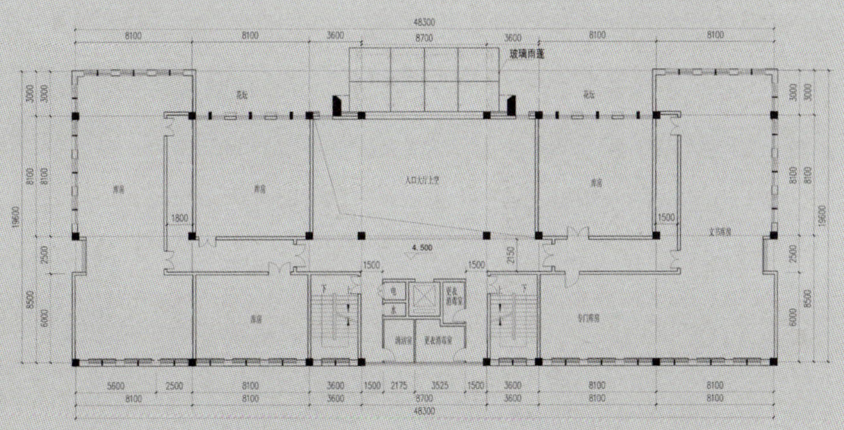
二层平面图 / Second Floor Plan

占地面积：972m²
建筑面积：4431m²
库房面积：1772m²
层　　数：5层
设计单位：深圳市市政设计研究院有限公司
建成时间：2014年10月

Site area: 972m²
Floor area: 4,431m²
Repository area: 1,772m²
Number of floors: 5 floors
Time of completion: October 2014

武宁县档案馆成立于1959年，现有纸质档案254673卷、75251件，照片档案1432张，声像267盘（件），实物档案190件，资料35080册。县档案馆于2015年被江西省档案局评为省一级档案馆。

武宁县档案馆新馆位于县城沙田新区中心地段，紧邻城市建设规划展览馆、博物馆、大剧院、县市民服务中心及各机关单位办公大楼，选址安全、便民。新馆于2012年3月开工建设，2014年10月建成投入使用。

建成后的武宁县档案馆功能齐全、布局合理。建筑一楼为对外服务用房，包含入口大厅、展览及文件阅览等公共服务功能；二、三楼为档案库房，单独设置，方便工作人员进行管理查阅；四楼为业务技术用房；五楼为办公用房。馆内各类设施设备完善，配备了高压细水雾灭火系统、视频监控系统、门禁及声光报警系统以及中央空调、除湿机等。

整个建筑造型严肃统一，彰显了文化性的设计理念，符合当代档案馆建筑的属性。

1 档案馆外景 / The Archives Exterior

Established in 1959, Wuning Archives has 254,673 files and 75,251 items of paper files, 1,432 photo archives, 267 (pieces of) audio/video disks, 190 material objects, and 35,080 volumes of documentary archives. The Archives was awarded as Provincial class-1 Archives by Jiangxi Provincial Archives Administration in 2015.

The new building is located in the core area of Shatian New District of the county. It is close to the Urban Construction Planning Exhibition Hall, Museum, Grand Theatre, the County's Citizen Service Center and office buildings of various agencies as the location is safe and convenient. The construction of the new building was started in March 2012 and completed and put into use in October 2014.

The new Archives provides complete functions and has a reasonable layout. On 1F are rooms for public services, including the Entrance Hall, exhibitions and document reading. On 2F and 3F are Archives Repositories, which are separately arranged to make it easier for the staff to manage and check, on 4F floor are business and technical rooms, and on 5F are office rooms. The Archives is well equipped, with high-pressure water mist fire extinguishing system, video surveillance system, access control and sound/light alarm system, as well as central air conditioning, dehumidifiers, etc.

The building's overall architectural style is solemn and unified, highlighting the cultural design concepts and adapting to its attributes as a contemporary Archives.

2
3 | 4
5 | 6

2 档案查阅中心 / Archives Reference Center
3 入口大厅 / Entrance Lobby
4 会议室 / Conference Room
5 特藏室 / Special Collection Room
6 库房 / Repository

修水县档案馆

Xiushui Archives

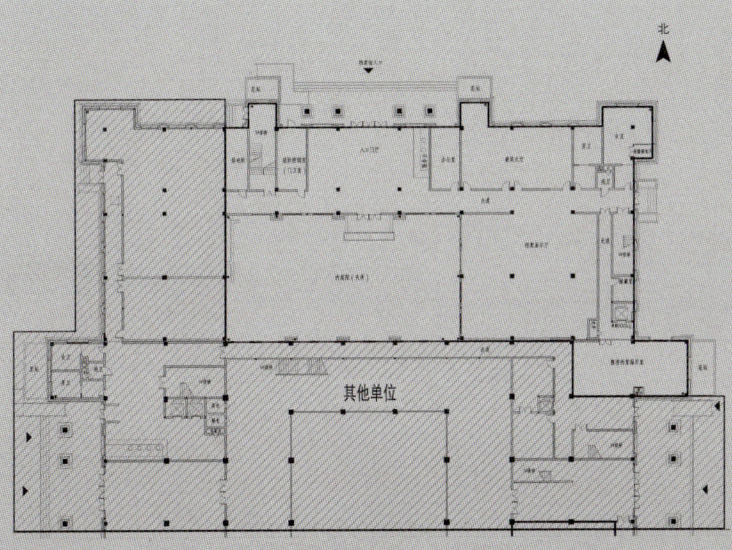

一层平面图 / Ground Floor Plan

占地面积： 1814m²
建筑面积： 5329m²
库房面积： 2200m²
层　　数： 3 层
设计单位： 浙江东华规划建筑园林设计有限公司
建成时间： 2013 年 4 月

Site area: 1,814m²
Floor area: 5,329m²
Repository area: 2,200m²
Number of floors: 3 floors
Time of completion: April 2013

修水县档案馆成立于1959年，馆藏档案160个全宗，18.7万余卷（件、册）。近年来，县档案馆大力加强档案资源建设、不断夯实档案基础业务，积极推进信息化建设，扎实做好档案开发利用工作，2017年被江西省档案局授予省一级档案馆称号。

修水县档案馆新馆位于新城良塘江渡大道，与城市展示馆、图书馆、博物馆四馆合建但相对独立，于2011年10月开工建设，2013年4月建成投入使用。

新馆选址合理、环境优美、设计新颖，功能布局合理、配套设备齐全，建筑呈分布式布局，设有独立的展厅和服务空间。展览空间设于一层，方便观众到达，各出入口均设有无障碍设施。馆内配备了高压细水雾灭火系统、温湿度自动调控系统、防盗报警监控系统以及中央空调、密集架等。

建筑造型古朴，结构严谨，院落式的布局使整个建筑活泼生动，呈现开放性的特征，充分体现了档案馆建筑的人文特征。

1 档案馆外景 / The Archives Exterior

Established in 1959, Xiushui Archives holds 160 fonds and more than 187,000 files (items and volumes) of archives. In recent years, the Archives has been vigorously improving archival resources, keeping consolidating basic archival business, actively promoting the construction of information technologies, and practically developing and utilizing archives. In 2017, it was awarded as Provincial class-1 Archives by Jiangxi Provincial Archives Administration.

Located on Liangtang Jiangdu Avenue in the new urban area of Xiushui County, The new building of Xiushui Archives was built together with the city's Exhibition Hall, Library and Museum while it was kept quite separated. Its construction was started in October 2011, and completed and put into use in April 2013.

The new building has an ideal site, beautiful environment, and novel design. Its functions are reasonable arranged, with completely equipped facilities. The building is in a distributed layout, with independent exhibition halls and service spaces. The exhibition space is on 1F for convenient accessing. Access-free facilities are available at all entrances. Inside it is equipped with high-pressure water mist fire extinguishing system, automatic temperature/ humidity control system, anti-theft alarm monitoring system, central air conditioners and compact shelving.

The architectural style is simple and rigorous. The layout of courtyard makes the entire building lively and vivid, fully reflects the openness and humanistic characteristics of an Archives.

2
3 | 4
5

2 入口大厅 / Entrance Lobby
3 库房 / Repository
4 档案查阅大厅 / Archives Reference Hall
5 展厅 / Exhibition Hall

睢县档案馆

Suixian Archives

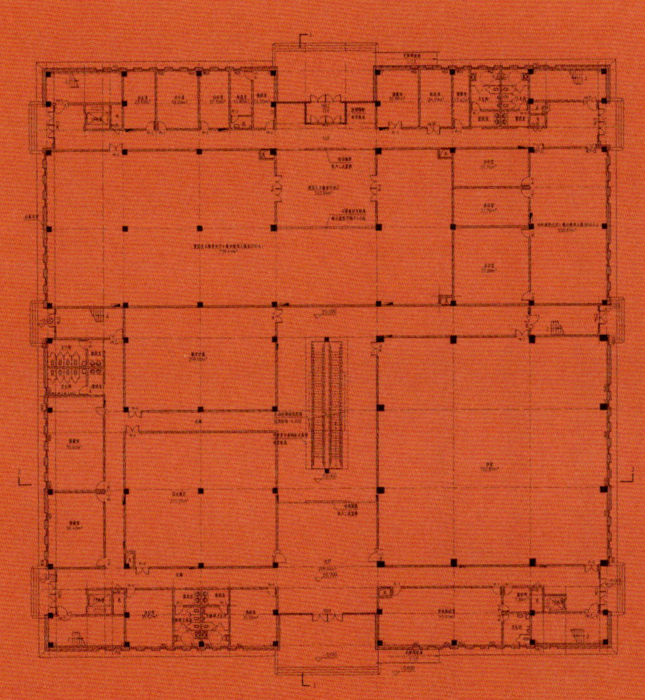

一层平面图 / Ground Floor Plan

占地面积： 2150m^2
建筑面积： 5000m^2
库房面积： 1646m^2
层　　数： 3 层
设计单位： 河南省建筑设计研究院有限公司
建成时间： 2017 年 7 月

Site area: 2,150m^2
Floor area: 5,000m^2
Repository area: 1,646m^2
Number of floors: 3 floors
Time of completion: July 2017

睢县档案馆始建 1958 年，目前馆藏档案和资料 10 万余卷（册），包括纸质、音像、照片等各种载体。

睢县档案馆新馆位于睢县风景优美的商务中心区内，与城市展览馆合建，于 2017 年 7 月建成投入使用，是全县档案资料的永久保存基地、重要的爱国主义教育基地、社会各界利用研究档案史料的中心。

新馆建筑共 3 层，功能齐全，设备先进。馆内设有库房区、办公区、档案利用服务中心、政府信息公开中心、电子文件管理中心等功能。其中库房面积 1646m²，可以满足近 30 年的档案库存量。《档案记忆》展区位于一层入口处，面积约 900m²，分为“厚重睢县、红色睢县、魅力睢县”三部分。整体布局按照“依托馆藏、立足本地、注重特色、教育为本”的思路，运用多媒体互动手段综合展示了睢县悠久的历史，改革开放以来经济、政治、文化取得的成就。

新馆建筑外观为完整的方形几何体，入口空间内凹处理，立面横竖条形的开窗，像是一段段睢县发展之路，共同勾勒出记载睢县文化的墙面符号，告诉人们睢县是怎样从古至今一路走到现在，又将如何走向未来。

1 档案馆外景 / The Archives Exterior

Initially built in 1958, Suixian Archives has a collection of more than 100,000 files (volumes) of archives and documents in all types of carriers like paper, audio and video, and photos.

The new Archives is located in a business center with splendid landscape in Suixian County. Built together with the City Exhibition Hall, it was completed and put into use in July 2017. It is the permanent base for preserving archives of the county and for patriotism education, as well as a center where all walks of life can access and use the archives and historical documents.

With three floors, the new building has complete functions and advanced equipments. There are the repository area, office area, archive access and use service center, government information disclosure center, electronic document management center, etc. The repositories are in 1,646m^2, which can keep archives coming in nearly the next 30 years. The exhibition area "Memory of Archives" is at the entrance on 1F in about 900m^2, which comprises three parts, namely Profound Suixian County, Red Suixian County, and Charming Suixian County. The overall layout is based on the idea "rely on the collections, based upon local resources, pay attention to characteristics, and regard education as the basis". It uses multimedia as the means of interaction to comprehensively show the long history of the County and the achievements in the fields of economy, politics and culture since the reform and opening-up.

The appearance of the new building is a complete square-shaped geometry. The entrance space is a concave, with windows in horizontal and vertical stripes on the facade being like sections of road on which the County seeks its development. They are the symbols on the wall that keep a record of its culture, and tell the people how has the County made its journey to the present from the past, and where will it head for.

2
3
4 I 5

2 展厅 / Exhibition Hall
3 档案查阅室 / Archives Reference Room
4 一层局部 / GF Local View
5 库房 / Repository

The Hongan County Archives was established in December 1958. With a collection of 215 archive fonds and more than 290,000 files (volumes) of archival holdings, it is a national class-1 general archives.

The new site of Hongan Archives is a single building located at the Cultural Center on Hongan Avenue. The construction was started in November 2011 and in December 2012 it was completed with a county-level national general archives construction site meeting was held. In September 2013, the Archives was moved to the new building and opened to the public.

The new building was built against the mountain. With the topography and terrain taken into consideration, the entrance space rises up through the flight of steps. The unique cylindrical body highlights the core space of the exhibition area and sets off the importance of Hongan Archives. When we look from above, the exhibition area in the front of the new building is like a "tam-tam (copper gong)", and the repositories and office area are like a "big knife", which imply "Huangma Uprising" and "a strike of a gong gathered 480,000 people" and subtly incorporate the local revolutionary history into the architectural design, highlighting the historical and cultural elements of Hongan Revolution.

The new building has clear streamlines and explicit functional divisions, namely the Exhibition Area, Repository Area, and Office and Technical Rooms. The Exhibition Area has two floors in full height with the roof made of transparent glasses. 1F is a public space providing the hallway, exhibition, public services among other functions. The circular corridor on 1F is the exhibition area, where columns are used to separate this area and the hallway. The straight flight stairs on both sides vertically guide people to the circular exhibition corridor on the mezzanine. The gray space between the straight flight stairs on both sides, leads to the office area, so that the two areas are both independent and interconnected. Functional rooms and facilities are fully equipped, facilitating security management of the archives and repositories, and keeping the workflows in order.

Hubei Province

旧馆 Old

New 新馆

2
3 | 4
5 | 6

2 展厅 1 / Exhibition Hall 1
3 展厅 2 / Exhibition Hall 2
4 办公区 / Office Area
5 查阅大厅 1 / Reference Hall 1
6 查阅大厅 2 / Reference Hall 2

黄梅县档案馆

Huangmei Archives

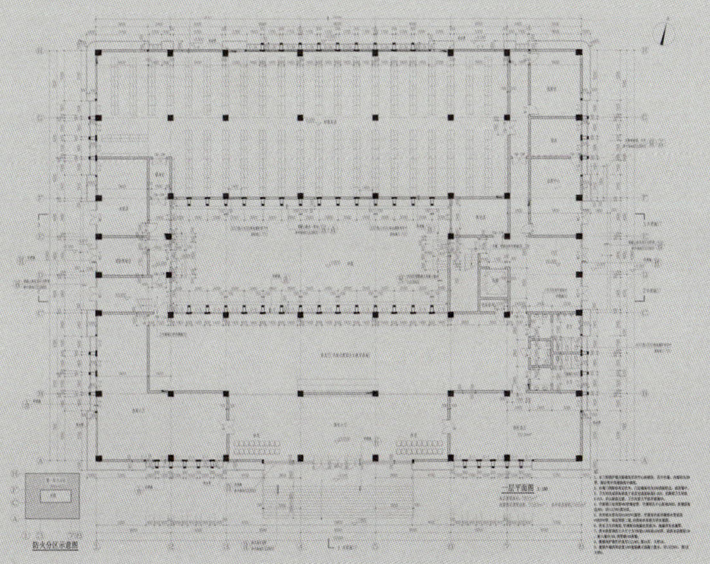

一层平面图 / Ground Floor Plan

占地面积： 5788m²
建筑面积： 5589m²
库房面积： 2166m²
层　　数： 3 层
设计单位： 中南建筑设计院
建成时间： 2016 年 7 月

Site area: 5,788m²
Floor area: 5,589m²
Repository area: 2,166m²
Number of floors: 3 floors
Time of completion: July 2016

黄梅县档案馆成立于1959年11月，馆藏168个全宗、档案资料18万卷（册），2017年被评为湖北省最美档案馆。

新馆占地面积5788m²，平面采用“回”字形，功能齐全，布局合理。“回”字形形成的内院划分档案库房区、办公区和展览区，内院以南为通高服务大厅、文件查阅中心、陈列馆、档案业务、技术用房及办公用房，内院以北为档案库房。

建筑造型端庄大气，虚实结合，两道回形石材为实，之间竖向格栅及玻璃窗为虚，入口处以大面玻璃幕墙既强调出入口核心功能，又与内部通高服务大厅相衬，外部造型设计与内部空间布局极好的对应统一，赋予服务大厅更多的开放性质，展现当代档案馆的开放精神。

技术方面，新馆安装有门禁、视频监控、恒温恒湿、微环境、高压细水雾自动消防、智能照明等系统；配有807组智能密集架，采用RFID电子标签技术；展厅采用光影技术，全方位展示出全县政治、经济、军事、文化、社会、党建等方面的发展历程。

Huangmei Archives was established in November 1959. With a collection of 168 fonds, and 180,000 files (volumes) of archival holdings.

In 2017, it was awarded "The Most Beautiful Archives of Hubei Province".

The new building covers an area of 5,788 square meters in the shape of the Chinese character " 回 ", providing complete functions with a reasonable layout. The inner courtyard is divided into the Archival Repository Area, Office Area and Exhibition Area. In the south of the courtyard are the full-height Service Hall, Document Reference Center, Exhibition Hall, Archives Service Room, Technical Room and Office Space. In the north is the Archival Repository.

The building adopts a dignified, magnificent style which combines virtuality with reality : two courses of square stones represent the reality and the vertical grilles and the glass windows in between resemble the virtuality. The large-sized glass curtain wall at the entrance not only emphasizes the core function of the entrance, but also corresponds to the well-hole style Service Hall inside. The external design and the layout of interior spaces are in perfect uniform, offering more openness to the Service Hall and showing the spirit openness that a contemporary archives building should have.

The new Archives employs high technologies including access control, video surveillance, constant temperature and humidity, micro-environment, automatic high-pressure water mist for fire protection, intelligent lighting and other systems. It is equipped with 807 intelligent compact shelving, and adopts RFID electronic tag technology. In the Exhibition Hall, light and shadow technology is applied to display the development process of the county's politics, economy, military, culture, society and party building, etc.

1 档案馆外景 / The Archives Exterior

2 | 3
4
5

2 档案查阅中心 / Archives Reference Center
3 入口大厅 / Entrance Lobby
4 展厅 / Exhibition Hall
5 库房 / Repository

建始县档案馆

Jianshi Archives

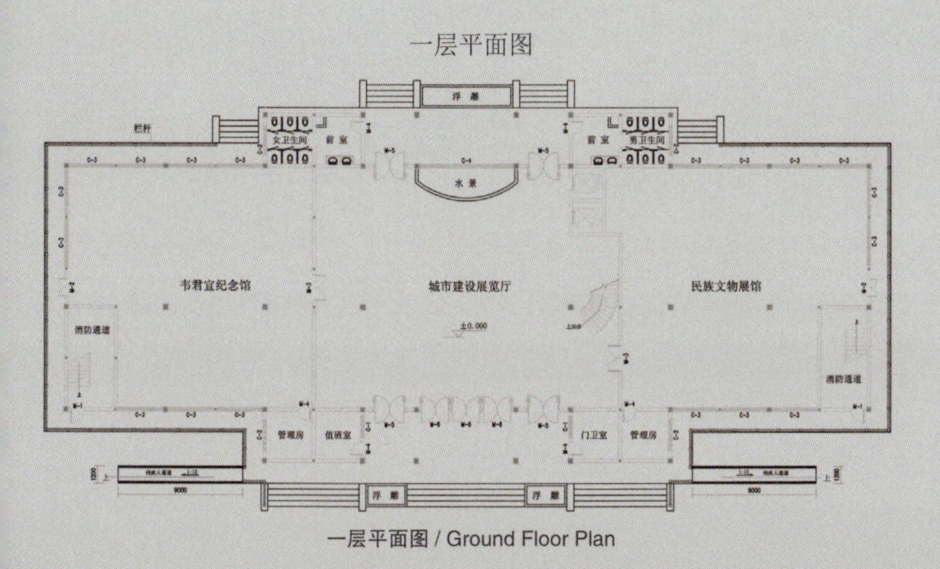

一层平面图 / Ground Floor Plan

占地面积： 1600m²
建筑面积： 6304m²
库房面积： 2400m²
层　　数： 7 层
设计单位： 湖北省大冶市中鲁古建园林设计公司
建成时间： 2012 年 6 月

Site area: 1,600m²
Floor area: 6,304m²
Repository area: 2,400m²
Number of floors: 7 floors
Time of completion: June 2012

建始县档案馆成立于1959年3月。馆藏124个全宗，档案资料30万余卷（册），是国家一级综合档案馆。档案馆于2016年被评为湖北省中小学生档案教育实践基地，2017年被评为湖北省最美档案馆。

建始县档案馆新馆位于朝阳大道，紧邻民族风情街，地处城区中心地带，交通便利；四周绿树成荫，花草密布，景色秀美。

档案馆主体为轴线对称的仿古风格建筑，雕栏画栋、翘檐塔顶，充分体现地域风格和民族特色，并在功能上设置民族文物展厅与民族文化活动中心，增强建筑功能与形式的统一性，赋予档案馆新馆民族文化的内涵。

新馆底层设置城市建设展览厅、韦君宜纪念馆及民族文物展馆，以多元的展览功能展现出较为开放的建筑氛围。新馆门厅空间与北侧次入口以水景相隔，既限定出门厅空间，又与北侧次入口保证了一定的连通性；门厅右侧弧形楼梯具有较强的垂直引导性，将人流引入档案馆功能区，二～六层档案馆功能区分区明确，各层西侧主要分布空间较大的图书库、档案室、资料室、大排练厅等功能，东侧主要分布空间较小的会议、办公、接待等功能。新馆顶层设置一大型民族文化活动中心。整栋建筑以“展览——档案——活动”为竖向功能分布，功能清晰、布局合理。

新馆采用电子钥匙、防紫外线灯管和新型防火材料，建有指纹识别、视频监控、温湿度调控、消防监测和灭火系统、档案信息化管理和安防系统等现代化管理设备。

1 档案馆外景 / The Archives Exterior

Jianshi Archives was established in March 1959. With 124 fonds and more than 300,000 files (volumes) of archives in its collection, it is a national class-1 general archives. In 2016, the Archives was named "Archives Education Practice Base for Primary and Secondary School Students of Hubei Province" and in 2017 "The Most Beautiful Archives of Hubei Province".

Located in Chaoyang Avenue, the Archives' new building is close to the National Customs Street, and in the core area of the city with convenient transportation. It is surrounded by trees, flowers and grass, enjoying beautiful sceneries.

Symmetric to its axis, the main body is an antique style building decorated with carved columns and spires. It shows the local architectural style and folk characteristics. It has a National Cultural Relics Exhibition Hall and a National Cultural Activity Center, enhancing the unity of architectural functions and forms, embodying connotations of nationality culture.

On 1F there are City Construction Exhibition Hall, Wei Junyi Memorial Hall and the National Cultural Relics Exhibition Hall to show a relatively open architectural atmosphere with diversified exhibition functions. The hallway of the new Archives is separated from the secondary entrance on the north with water landscape, which not only defines the space of the hallway, but also guarantees connectivity with the secondary entrance on the north side; the curved staircase on the right side of the hallway, with obvious vertical guidance, leads the people to enter the functional area of the building. The functional areas from 2F to 6F are also clearly defined. On the west side of the building are mainly functions with large space like Library, Archives Repositories, Documentary Archives Room, and Large Rehearsal Hall. On the east side, there are mainly functional areas with small space for meetings, office, reception, etc. A large-scale national cultural activity center is arranged on the top floor of the new building. The whole building is distributed vertically by the functions "exhibition - archives - activity". The functions are clearly defined with a reasonable layout.

The new building adopts electronic keys, UV protection lamps and fireproof materials, and is equipped with modern management equipments such as fingerprint recognition, video surveillance, temperature and humidity control, fire monitoring and fire fighting system, archives information management and security systems.

2 目录室 / Catalogue Room
3 展厅 1 / Exhibition Hall 1
4 展厅 2 / Exhibition Hall 2
5 展厅 3 / Exhibition Hall 3
6 档案查阅室 / Archives Reference Room

利川市档案馆
Lichuan Archives

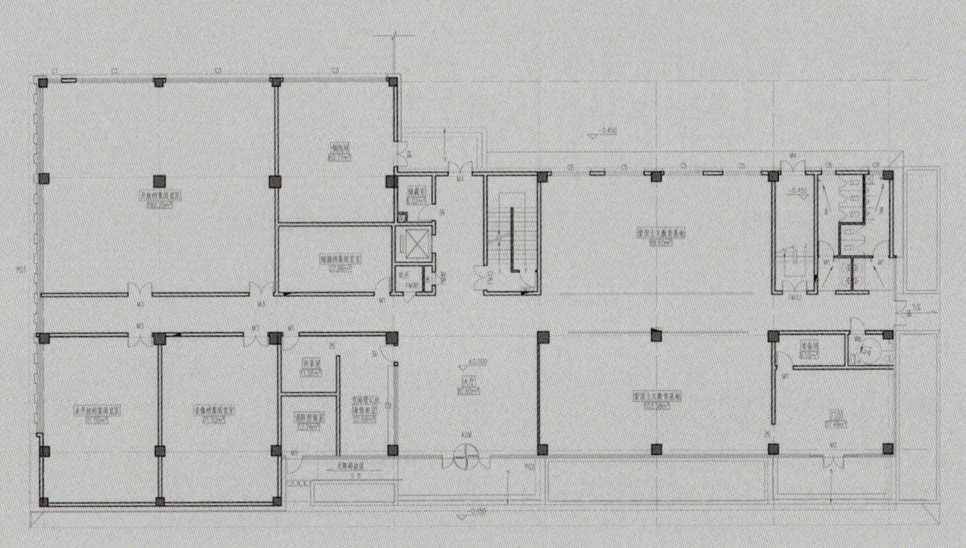

一层平面图 / Ground Floor Plan

占地面积： 8000m²
建筑面积： 6775m²
库房面积： 2200m²
层　　数： 地上6层，地下1层
设计单位： 华中科技大学建筑设计研究院
建成时间： 2017年12月

Site area: 8,000m²
Floor area: 6,775m²
Repository area: 2,200m²
Number of floors: ↑ 6 floors and ↓ 1 level
Time of completion: December 2017

利川市档案馆成立于1958年，馆藏201个全宗，档案资料38万卷（册），是国家二级综合档案馆。

利川市档案馆新馆位于利川市城南新区，是党政机关和公共文化设施集中区，城市基础设施完善，交通便利，周边环境优美。

利川市档案馆为6层独立建筑。建筑立足土家族苗族的地域特征，结合后现代乡土主义风格，整体造型体量咬合与穿插。各个立面虚实结合，玻璃与石材结合出灰空间的元素，简洁统一。正立面两处虚空间的设计，门厅处的灰空间以及以颜色突显的大门暗示了建筑的主入口空间，中间大片通透的玻璃幕墙，激活建筑内部空间与外部环境的视线交流，同时对内部空间的功能进行暗示。整个立面虚实结合，“实中有虚，虚中有实”，展现的恰到好处。

档案馆新馆平面布局主要以门厅为界分为东西两部分，西侧为空间较大的档案阅览区、档案库房及会议区，东侧为面积较小的行政办公、技术办公及资料室等。库房垂直分布，每层配有相应的技术办公与行政办公，以东西向的走廊相通，便捷档案馆的管理与运作，提高效率。

同时，新馆配置有中央空调系统、自动气体消防和喷淋消防系统、库房温湿度自动调控系统、库房空气自动消毒系统、库房门禁系统、视频监控和远红外报警系统等设施设备，保障档案馆安全高效的运行。

Set up in 1958, Lichuan Archives has a collection of 201 fonds, and more than 380,000 files (volumes) of documentary archives. It is a national class-2 general archives.

The new building of Lichuan Archives is located in Chengnan New District, Lichuan City, where there are many Party and government organs and public cultural facilities with well-developed infrastructure, convenient transportation and beautiful environment.

The Archives is a six-storey independent building. Based on the regional characteristics of the Tujia and Miao nationalities and combined with the post-modern vernacularism style. The building has masses that are interspersed with each other, with the facade combining virtuality with reality, and glasses used together with stones to create elements of gray spaces, making the overall appearance simple and unified. The design of the virtual spaces at the front facade, the gray space at the hallway and the gate highlighted by colors imply the main entrance space of the building. The large transparent glass curtain wall in the middle activates visual communication between interior and the external environments of the building while functions inside the building are also implied so that the whole facade integrates virtuality with reality and properly shows such integration off.

The new Archives is mainly divided by the hallway into the east and the west parts. On the west side are the Archive Reading Area, Archives Repositories and Conference Area while on the east side are Administrative Office Area, Technical Office Rooms and Documentary Archives Rooms. The repositories are vertically distributed, with corresponding technical offices and administrative offices arranged on each floor, which are identified with the east-west corridor, in order to facilitate management and operation of Archives and improve efficiency.

Meanwhile the new building is equipped with central air conditioning system, automatic gas fire system and sprinkler system, repository temperature and humidity automatic control system, repository air automatic disinfection system, repository access control system, video surveillance and far infrared alarm system, and other facilities and equipments so as to guarantee safe and efficient operation of the Archives.

1 档案馆外景 / The Archives Exterior

2 | 3
4 | 5

2 档案查阅大厅 / Archives Reference Hall
3 展厅 1 / Exhibition Hall 1
4 展厅 2 / Exhibition Hall 2
5 库房 / Repository

松滋市档案馆

Songzi Archives

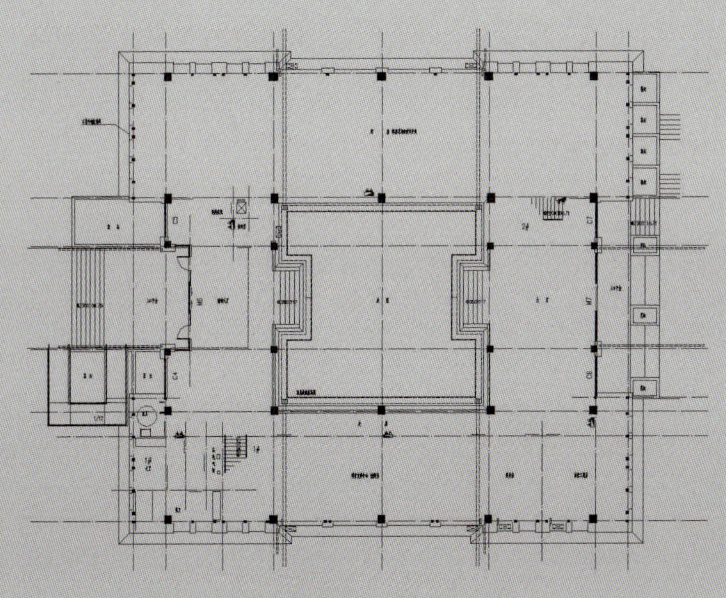

一层平面图 / Ground Floor Plan

占地面积： 1056m²
建筑面积： 4054m²
库房面积： 1488m²
层　　数： 4 层
设计单位： 武汉东艺建筑设计有限公司
建成时间： 2011 年 6 月

Site area: 1,056m²
Floor area: 4,054m²
Repository area: 1,488m²
Number of floors: 4 floors
Time of completion: June 2011

松滋市档案馆成立于1959年4月，馆藏135个全宗，档案资料12万余卷（册），为国家一级综合档案馆。近年来，被评为湖北中小学档案教育实践基地、湖北省红色记忆档案馆及湖北省最美档案馆。

松滋市档案馆新馆位于松滋市全新规划设计的市民中心广场，地理位置优越，交通便利，公共服务设施齐全，生态环境优美，开放式的园林格局尽显亲民理念。

建筑以“回”字形布局，营造的内部庭院与西侧湖面自然环境相呼应。档案馆新馆为4层单体建筑，外观造型中轴对称，宽屋檐与大坡面结合的设计，气势恢宏，展现楚文化的深厚底蕴以及档案馆的神圣地位。

“回”字形平面北侧为库房区，南侧为办公区，东西两侧底层以虚空间相连，门厅、内院、次门厅与西侧湖面景观之间形成明显的视线轴线通廊，同时，沿内院一侧布置走道，激活内院空间，增强馆内各层之间的视线交流。内院的营建及通高空间的营造，赋予新馆内部空间丰富的变化，展现新馆的活力。

Set up in April 1959, Songzi Archives has a collection of 135 archive fonds, and more than 120,000 files (volumes) of archival holdings. It is a national class-1 general archives. In recent years, it has been titled Archives Education Practice Base of Primary and Secondary Schools in Hubei, Red Memory Archives of Hubei and the Most Beautiful Archives of Hubei.

The new site of Songzi Archives is located at the center square of Songzi City that has been recently planned and designed. It enjoys a superb geographical location, convenient transportation, complete public service facilities and a beautiful ecological environment. The open gardening layout fully embodies the concept “intimacy to people”. The building takes the shape of Chinese character “ 回 ”. The courtyard designed inside the building works in concert with the natural environment of the lake in the west.

The new Archives is a four-story single building with an axisymmetric structure. Wide eaves in combination with large slopes produce magnificence, reflecting the profound heritage of the Chu Culture and the sacred status of the Archives.

In the north of the building is the Repository Area, and in the south is the Office Area. Spaces on 1F at both the east and west sides are connected with a virtual space. The entrance hall, inner courtyard, secondary hallway and the lake at the west side form an evident visual axis corridor. At the same time, the walkway arranged along one side of the courtyard enlivens the courtyard spaces, and enhances eye contact across floors in the building. The courtyard and full-height style provide the interior spaces with rich changes, demonstrating the vitality of the new building.

1 档案馆外景 / The Archives Exterior

2 | 3 | 4
5
6

2 会议室 / Conference Room
3 内庭院 / Inner Courtyard
4 展厅 / Exhibition Hall
5 查阅室 / Reference Room
6 一层中庭 / GF Atrium

咸宁市咸安区档案馆

Xian'an District Archives, Xianning

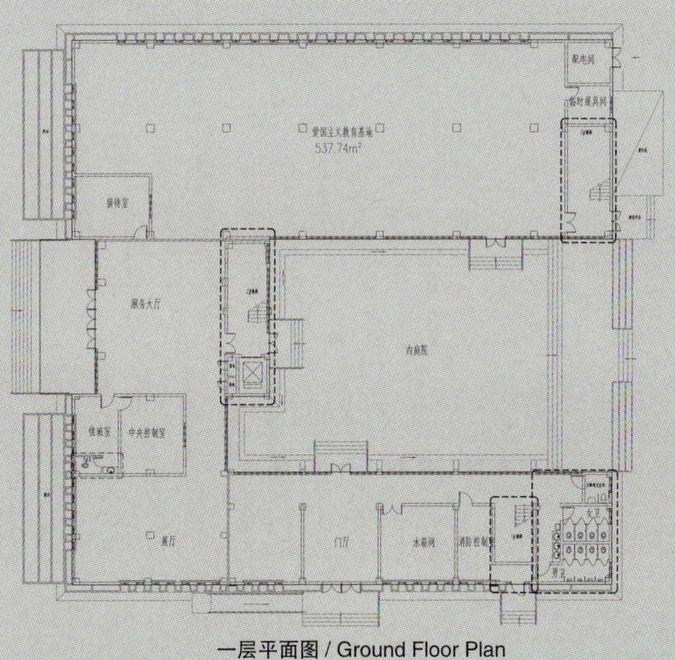

一层平面图 / Ground Floor Plan

占地面积： 10000m²
建筑面积： 4954m²
库房面积： 1833m²
层　　数： 4 层
设计单位： 湖北中江建筑设计有限公司
建成时间： 2017 年 12 月

Site area: 10,000m²
Floor area: 4,954m²
Repository area: 1,833m²
Number of floors: 4 floors
Time of completion: December 2017

咸宁市咸安区档案馆新馆于 2015 年 5 月开工建设，2017 年 12 月建成。新馆位于湖北省咸宁市咸安区银桂路中段东侧，处于市政府、区政府中心地带，地理位置优越，绿化景观优美。

咸宁市咸安区档案馆新馆为地上四层的单体建筑，新馆平面采用 C 形布局，南侧为技术及行政办公用房，北侧为档案库房，东侧为对外服务大厅，三者虚实相间，半围合内院中设置景观小品及活动场地，与入口广场形成轴线关系；空间形态上，形成了内外相间的空间变化关系。

新馆平面功能上，将对外服务及展览等开敞功能置于底层，展厅中常设的“天下咸安”大型爱国主义教育展览，具有鲜明的红色文化特色；档案库房及技术办公位于二～四层，设特藏库，文书库房、林权档案专库、人口普查档案专库、科技档案专库、资料专库、防磁库等，库房中装有温湿度、消毒、监控、消防自动控制系统。

建筑立面多采用竖向白色的大理石与玻璃窗相结合，竖向线条明显，营造出一种向上的动势。同时，立面由横向玻璃与篆书馆名形成水平虚空间分割，虚实结合，丰富整体造型，并分隔内部竖向空间。

新馆建筑设计中贯穿“咸安印记”的理念，以“印”为信物，档案为凭据的标识，彰显独特的区域文化与档案文化内涵。

1 档案馆外景 / The Archives Exterior

The construction of the new site of Xian'an District Archives in Xianning was started in May 2015 and completed in December 2017. The new Archives located on the east side in the middle section of Yingui Road, Xian'an District, Hubei Province. In the center of the municipal government and district government buildings, it enjoys a superior geographical position and beautiful green landscape.

The new Archives is a four-story single building It adopts a C-shaped layout. Its south part accommodates technical rooms and administrative offices, on the north side are the Archival Repositories, and in the east is the Public Service Hall. These three parts are arranged in a way to reflect both virtuality and reality. In a semi-enclosed courtyard there are landscape sketches and event spaces, on the same axis with the square at the entrance, and echoing with each other from inside and outside in terms of spatial relationship.

In the functional layout, open functions like public services and exhibitions are on 1F. The large-scale patriotism education exhibition "Tianxia Xian'an" is constantly on display in the Exhibition Hall, which has distinctive cultural characteristics. Archival Repositories and Technical Rooms are on 2F to 4F, including the Special Collection Room, Administrative Archival Repository, Special Forest Right Repository, Special Census Repository, Special Science and Technology Repository, Special Documentary Repository, Antimagnetic Repository, etc. Automatic temperature and humidity, disinfection, monitoring and fire control systems are furnished in the rooms.

The building's facade is mostly constructed with white marbles and glass windows vertically, forming distinct vertical lines and creating an upward momentum. At the same time, the horizontal glasses and the archives name in seal script seems like a virtual horizontal sword that divides the facade to combine virtuality with reality, enriching the overall appearance and partitioning the internal spaces vertically.

The concept of "Xian'an Seal" runs through the new building's design, where "seal" is deemed as the token, and archives as the sign of evidence. It highlights manifest unique regional culture and cultural connotations of archiving.

2 | 3
4 | 5
6

2 多功能厅 / Multi-purpose Hall
3 一层大厅 / GF Lobby
4 中控室 / Central Control Room
5 外立面局部 / Partial Façade
6 裱糊室 / Papering Room

阳新县档案馆

Yangxin Archives

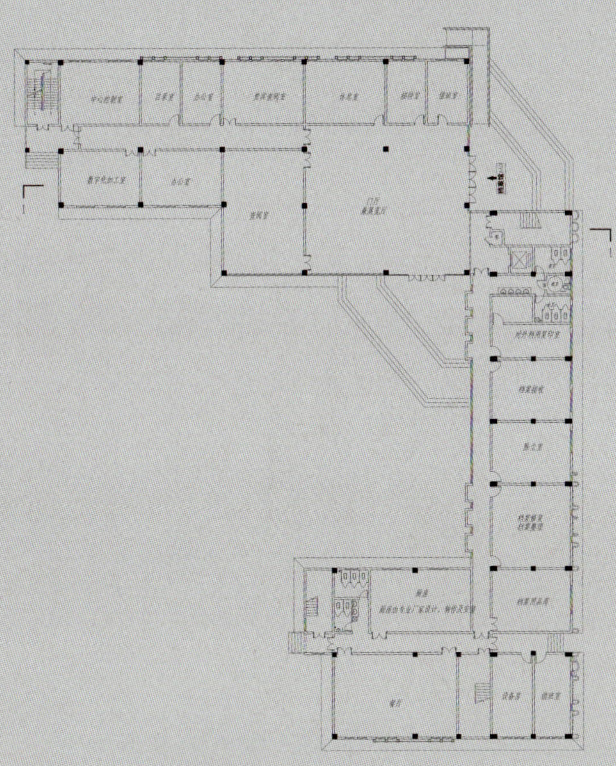

一层平面图 / Ground Floor Plan

占地面积：6670m²
建筑面积：5579m²
库房面积：2095m²
层　　数：6 层
设计单位：武汉宏图阁建筑设计有限公司
建成时间：2015 年 7 月

Site area: 6,670m²
Floor area: 5,579m²
Repository area: 2,095m²
Number of floors: 6 floors
Time of completion: July 2015

阳新县档案馆馆藏档案资料 17 万余卷（册），于 2017 年被评为湖北省红色记忆档案馆和湖北省最美档案馆。

阳新县档案馆新馆坐落于阳新县城东新区莲花湖畔，地理位置优越，周围绿化环境优美，公众服务设施齐全。

档案馆新馆在矩形的场地中采用 C 形布局，由此将场地划分出一个内部庭院，营造出较好的景观环境，同时提供给建筑内侧的功能房间一定的景观视觉。新馆主体建筑外观借用书侧脊形式，层层竖向线条排列，既表现出建筑向上的动势，展现建筑的挺拔，又体现档案牍册的文化内涵。

新馆主入口位于建筑东北角，处于对外服务和行政办公与技术办公和档案库房的交界处，即入口门厅右侧为对外服务和行政办公用房，行政办公与对外服务用房竖向交叉布置，提高档案馆服务工作效率；门厅左侧为技术办公和档案库房用房，技术办公在下，库房在上，利于办公人员之间的交流，同时，于建筑外立面而言，办公相对于库房需要更好的采光，外立面“下虚上实”，立面设计彰显建筑厚实稳重，对光照的适量控制，保障了档案库房的稳定性。

1 档案馆外景 / The Archives Exterior

Yangxin Archives holds more than 170,000 files (volumes) of archives. In 2017, it was awarded "Red Memory Archives of Hubei Province" and "The Most Beautiful Archives of Hubei Province".

Located in the bank of Lianhua Lake in Chengdong New District, Yangxin County, the new building of Yangxin Archives enjoys a superior geographical location, beautiful environment, and well-equipped public service facilities.

The new Archives adopts a C-shaped layout in a rectangular site, which forms a courtyard inside, which creates more beautiful landscape, and at the same time offers the functional rooms at the internal side of the building with certain views. The main building's appearance borrows the form of a book spine, and is arranged in vertical lines floor by floor, not only showing an upward momentum and uprightness of the building, but also reflecting the cultural connotation of archives and books.

The main entrance is at the northeast corner of the building, which is at the junction of the public service and administrative office and technical rooms and archives repositories, that is, on the right side of the entrance hall are the public service and administrative office rooms. The administrative office and public service rooms are intersected vertically to improve the efficiency of the Archives. On the left side of the entrance hall are the technical rooms and archives repositories. The technical office rooms are below the repositories, which makes it easier for the staff to communicate with each other. At the same time, as the offices need better lighting compared to the repositories, the principle "empty lower side and solid upper side" followed for the facade highlights the solidness and stability of the facade design while the lighting can be controlled properly, so as to guarantee stable operation of the repositories.

2 | 3
4
5 | 6

2 内庭院 / Inner Courtyard
3 档案查阅大厅 1 / Archives Reference Hall 1
4 展厅 / Exhibition Hall
5 档案查阅大厅 2 / Archives Reference Hall 2
6 多功能厅 / Multi-purpose Hall

浏阳市档案馆

Liuyang Archives

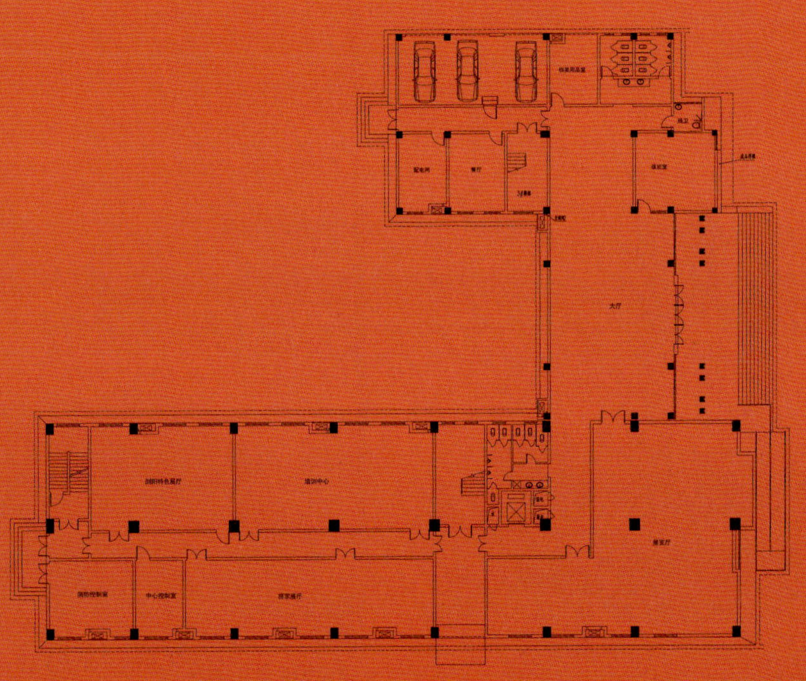
一层平面图 / Ground Floor Plan

占地面积： 7630m²
建筑面积： 7081m²
库房面积： 3015m²
层　　数： 主楼 6 层，配楼 4 层，地下 1 层
设计单位： 湖南方圆建筑工程设计有限公司
建成时间： 2012 年 12 月

Site area: 7,630m²
Floor area: 7,081m²
Repository area: 3,015m²
Number of floors: 6 floors for the main building, 4 floors for the wing and ↓ 1 level
Time of completion: December 2012

浏阳市档案馆建于 1958 年 12 月，目前馆藏档案 35 万余卷（册），2018 年被评为长沙市爱国主义教育基地。

浏阳市档案馆新馆位于湖南省浏阳市复兴中路 3 号，作为仅供档案馆使用的独立建筑，其四周以围墙围合出相对独立的空间，为前来参观的各界人士营造出相对安静舒适的展览空间。

新馆平面呈 U 形布局，整体功能布局分明，流线清晰。新馆以入口大厅衔接南北两侧不同功能用房，南侧的档案楼与北侧的办公楼，既相互联系，又有一定的独立性，保证了外来参观人员与内部工作人员的互不干扰。

在保证与行政中心周边建筑群总体协调的前提下，新馆力求体现档案馆作为历史文化建筑所具有的深邃厚重、简洁古朴的风格。因此，在立面设计上，外墙均采用白色大理石，以求两栋楼房协调统一融为一体。同时，在入口等重要之处设有地方特色的砖红色标志浮雕。白色大理石中点缀砖红色浮雕，端庄典雅中又延续有历史的记忆，就像在茫茫的时间长河中，档案馆替我们记录下历史的痕迹。

新馆一楼设有“浏阳记忆”展馆，包括古代展厅、近现代展厅、当代展厅、专题展厅及珍展室，面积 1200m^2，是对外展示浏阳形象的窗口和扉页，同时也是“知我浏阳，爱我浏阳”的爱国主义教育殿堂。

Built in December 1958, Liuyang Archives now has a collection of 350,000 files (volumes). It was awarded Patriotism Education Base of Changsha in 2018.

The new site of Liuyang Archives is located in No.3 Middle Fuxing Road, Liuyang, Hunan Province. It is an independent building used exclusively by the Archives, surrounded with enclosing walls to create relatively quiet, comfortable spaces for the visitors.

The new Archives features a U-shaped layout, with distinctive functional areas and clear streamlines. Functional rooms at the north and south sides of the new building are bridged by the entrance hall. The Archives Building on the south side and the Office Building on the north side are both interconnected and independent of each other, offering separate spaces for visitors and staff members.

While keeping the overall coordination with the buildings surrounding the administrative center, the new Archives' design aims to reflect its profoundness and primitive simplicity as a historical and cultural building. Therefore, white marbles are adopted for exterior walls on the facade, maintaining a unified appearance of the two buildings. Entrances and other important places are decorated with brick-red embossment with local features. These reliefs embellishing the white marbles resemble the historical memories continuing in elegance. It's just like the Archives which keeps our traces in the vast history.

The “Liuyang Memories” hall on 1F comprises of an Ancient Exhibition Room, Modern Exhibition Room, Contemporary Exhibition Room, Thematic Exhibition Room and Collection Exhibition Room, occupying a total area of 1200m^2. It shows the image of Liuyang to the outer world, and can also act as a classroom for patriotism education.

1 档案馆外景 / The Archives Exterior

浏陽市國家綜合檔案館

2 | 3
4 | 5
6

2 阅览室 / Reading Room
3 入口 / Entrance
4 库房 / Repository
5 展厅 / Exhibition Hall
6 档案查阅室 / Archives Reference Room

平江县档案馆

Pingjiang Archives

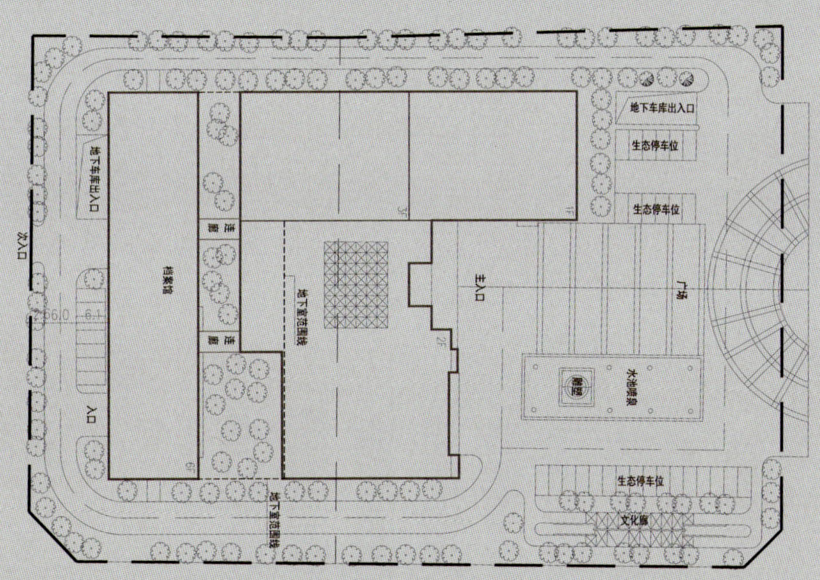

一层平面图 / Ground Floor Plan

占地面积： 13474m²
建筑面积： 16528m²
库房面积： 4700m²
层　　数： 主楼 6 层、配楼 3 层、地下 1 层
设计单位： 湖南省长沙华艺工程设计有限公司
建成时间： 2015 年 1 月

Site area: 13,474m²
Floor area: 16,528m²
Repository area: 4,700m²
Number of floors: 6 floors for the main building, 3 floors for the wing and ↓ 1 level
Time of completion: January 2015

平江县档案馆于 1958 年成立，原馆建成于二十世纪七十年代，现有馆藏档案资料 18.51 万卷（册）。平江县档案馆是岳阳市科普示范基地、岳阳市爱国主义教育基地，目前已成为平江县对外开放的重要窗口，是推介“绿富双赢新平江”的一张靓丽名片。

平江县档案馆新馆于 2013 年 2 月 26 日正式开工，与县规划馆和周令钊美术馆“三馆合一”，实现大型综合性档案馆建设。新馆分为主楼、裙楼及文化广场三大部分，主楼共六层，一楼主要是对外服务用房，二楼为办公和技术性用房，三～六楼为库房，裙楼为三层，设有展厅、报告厅、会议室等功能用房。裙房与主楼之间以两个连廊相连，既相对独立又互相联系，有效地避免了参观人员与工作人员活动流线的交叉，确保了良好的参观体验。

建筑立面处理上采用虚实结合的手法，通过体块的穿插，使得整个建筑在端庄典雅中又透露出一种蓬勃的朝气。馆内配置有先进的智能档案密集架、智能书车、细水雾喷淋和七氟丙烷气体灭火系统、智能监控防护系统。

平江县档案馆新馆的落成，有力地提升了档案工作的社会地位，为档案工作服务于全市经济社会的发展大局，推动平江县档案工作的进一步发展，打下了坚实的物质基础。

1 档案馆外景 / The Archives Exterior

Pingjiang Archives was established in 1958, and the original building was built in the 1970s. Now it has a collection of 185,100 files (volumes). The Archives has been titled Science Popularization Education Demonstration Base of Yueyang City, and Patriotism Education Base of Yueyang City. To Pingjiang County, it has become an important window to the outside world, and an impressive name card for promoting "new Pingjiang with achievements in both environment protection and economic development".

The construction project of Pingjiang Archives' new building was officially commenced on February 26, 2013. It is part of the large-scale comprehensive project "Three Museums in One", which also comprises the County Planning Museum and Zhoulingzhao Art Museum. The new building consists of three major parts, namely the main building, the podium building and the cultural square. The main building has six floors, with 1F mainly accommodating public service rooms, 2F for office and technical rooms, and 3F to 6F for repositories; functional rooms are arranged on the podium building with three floors, including Exhibition Hall, Lecture Hall, Conference Rooms, etc. Two corridors link the podium building with the main building to ensure they are relatively independent while interconnected, effectively avoiding the overlapping of visitor and staff streamlines and ensuring excellent visiting experience.

The building's facade is designed to combine reality with virtuality; intercrosses of masses are adopted to reflect dignity and elegance while incarnate vitalities at the same time. Equipped with advanced intelligent compact archives shelving, smart book trolleys, mist sprinklers and heptafluoropropane gas fire extinguishing system, and intelligent monitoring and protection system.

The new site of Pingjiang Archives has effectively enhanced the social status of archiving, and laid a solid foundation for archiving to serve the overall economic and social development of the city, and promoting further development of Pingjiang County's archiving.

展厅 1 / Exhibition Hall 1

入口大厅 / Entrance Lobby

展厅 2 / Exhibition Hall 2

湘潭县档案馆

Xiangtan Archives

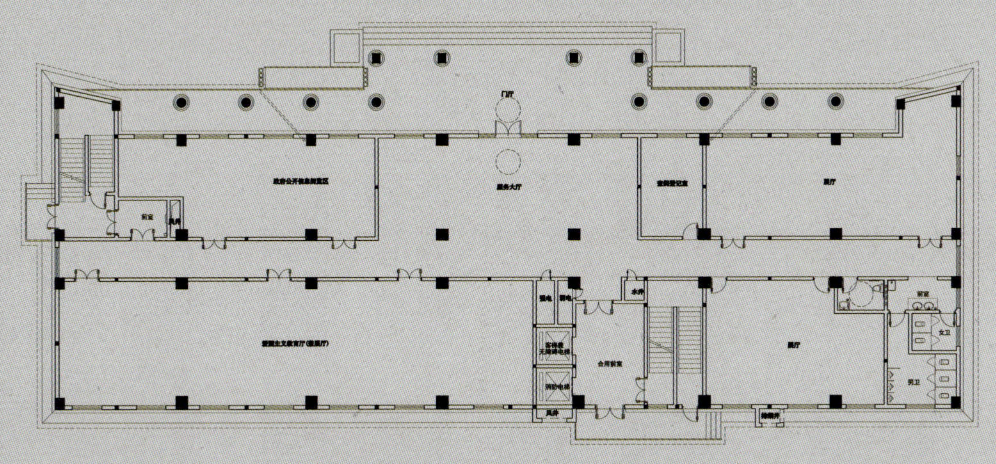

一层平面图 / Ground Floor Plan

占地面积：5420m²
建筑面积：13521m²
库房面积：6800m²
层　　数：主楼 10 层，地下 1 层
设计单位：湖南省湘潭市规划建筑设计院
建成时间：2017 年 1 月

Site area: 5,420m²
Floor area: 13,521m²
Repository area: 6,800m²
Number of floors: 10 floors =and 1 level
Time of completion: January 2017

湘潭县档案馆成立于1958年12月13日，现有馆藏档案50万卷册，其中国家重点档案26995卷册，还藏有毛泽东、彭德怀、齐白石等伟人、名人档案。

湘潭县档案馆新馆于2014年11月在湘潭天易经开区飞鸽路开工建设，按照档案馆功能定位和湘潭天易经开区建筑风格设计。建筑外立面采用三段式造型，一二层灰空间划分为第一段，三～十层规整的竖窗为第二段，顶部的实墙檐口为第三段；竖向以凸出的玻璃幕墙及门头彰显入口门厅的核心地位，两侧的竖向实墙及竖向窗间墙的突出，增强建筑竖向的动势，使得建筑更为挺拔。总体来说，新馆建筑造型规整统一，简洁大方，气势恢宏。

新馆一层、二层为主题展馆，全面介绍湘潭县社会经济发展和人文底蕴，展示湖湘文化和莲乡档案文化特色；入口门厅通高的设计，与室外挑廊灰空间及室内非通高空间形成了“室外空间——半室外灰空间——室内通高空间——室内非通高空间”的空间变化关系；三～九层为档案库房和办公会议用房，顶层设有一报告厅，可容纳150人左右，足以满足档案馆的会议工作需求。

1 档案馆外景 / The Archives Exterior

Established on December 13, 1958, Xiangtan Archives has a collection of 500,000 files, including 26,995 national key archival files. In the Archives' holdings, there are records about great persons and celebrities like Mao Zedong, Peng Dehuai and Qi Baishi.

Construction of Xiangtan Archives' new building was started in November 2014 on Feige Road in Xiangtan's Xiangyi Economic Development Zone. Based on the Archives' functional planning and the Zone's architectural style, a three-part facade is designed, with the gray spaces of 1F and 2F being the first part, regular vertical windows on 3F to 10F being the second part, and solid wall eaves on top being the third part; vertically, the protruding glass curtain wall and the lintel highlight the key position of the hallway; vertical stone walls and piers between windows, both projective, beef up the vertical momentum of the building and making it more towering and straight. In general, the new building takes a uniform, simple while graceful appearance which reflects its magnificent, grand manner.

1F and 2F of the building is the Subject Exhibition Hall, which comprehensively presents the social-economic development and cultural heritage of Xiangtan County, and displays the characteristics of Huxiang culture and Lianxiang archival culture; the entrance hallway is in a well-hole style, in combination with the outdoor overhanging corridor's gray spaces and indoor non full-height spaces forming the spatial relationship "outdoor space - semi-outdoor gray space - indoor full-height space - indoor non full-height space"; from 3F to 9F are the Archives Repositories and Office Conference Rooms, and on the top floor a Lecture Hall accommodating about 150 persons is provided to meet the relevant requirements.

2
3 | 4
5 | 6

2 展厅 / Exhibition Hall
3 档案阅览室 / Reading Room
4 档案整理室 / Archives Filing Room
5 中控室 / Central Control Room
6 报告厅 / Lecture Hall

荔浦县档案馆

Lipu Archives

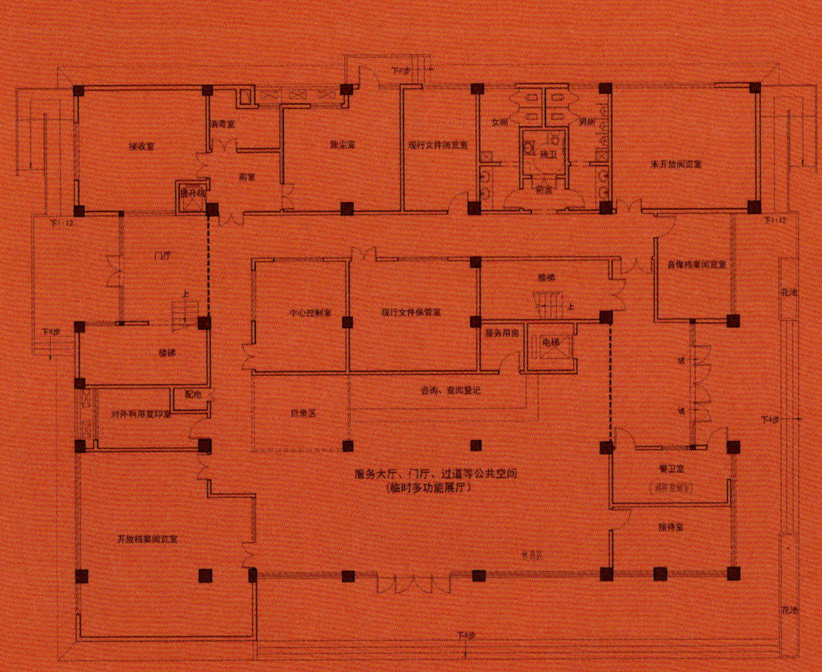

一层平面图 / Ground Floor Plan

占地面积： 1957m^2
建筑面积： 3665m^2
库房面积： 1102m^2
层　　数： 5 层
设计单位： 桂林市建筑设计研究院
建成时间： 2015 年 3 月

Site area: 1,957m^2
Floor area: 3,665m^2
Repository area: 1,102m^2
Number of floors: 5 floors
Time of completion: March 2015

旧馆 Old

New 新馆

荔浦县档案馆成立于 1958 年 12 月，目前馆藏档案 169 个全宗，共 46320 卷，馆藏资料 10218 册。

荔浦县档案馆新馆位于县城滨江南岸行政办公区，于 2012 年 8 月破土动工，2015 年 3 月建成投入使用。该项目荣获 2017 年度桂林市优秀工程勘察设计一等奖、广西优秀工程设计二等奖。

档案馆新馆设计采用“回”字形的总平面布局形态，中心宛如一枚中国印，整体形状似“典”字，寓意珍贵典籍档案藏于馆中。立面造型设计，体块穿插明显，虚实相间，西立面在门厅入口处营造灰空间，并在其屋面设计休息平台，激活档案馆的屋面空间；内凹的红色纹理格栅网，如同阳文的印章，展现中国文化博大精深的文化内涵；北立面突出的体块以竖向线条分割，类似古籍的书脊，暗示建筑与古籍相关的功能性质。

新馆一层平面类似于风车状，以 3 个主次入口将“回”字形外轮廓分成三段，分别对应展览区、技术办公区以及辅助服务区；二、三、四、五层以 C 形为平面形态，水平向布置档案库房、技术用房和行政办公用房，围合的空间形成景观独特的上人屋面。

档案馆新馆的设计，功能合理，造型新颖，可满足荔浦县未来 50 年档案事业发展需要。

Established in December 1958, Lipu Archives has a collection of 169 fonds, with 46,320 files and 10,218 volumes of documentary archives.

The new building of the Archives is located in the Administrative Office Area on the south bank of the county. Its construction project was commenced in August 2012, and completed in March 2015 for putting into use. The Project won the first prize of Excellent Engineering Survey and Design of Guilin 2017, and the second prize of Excellent Engineering Design of Guangxi.

The new Archives takes a general plan layout in the shape of the Chinese character " 回 ". The center part is just like a Chinese seal, and the overall form resembles the character " 典 ", which means "ancient books and records", to imply the precious holdings reserved in it. The facade is designed with evident intercrossed massing that combines reality with virtuality; the west facade produces a gray space at the entrance of hallway, with a rest platform designed on its roof to expand the Archives' roof spaces; the concave red texture grille is similar to an embossed seal, reflecting the profound cultural connotations of Chinese culture; the prominent massing of the north facade, as well as the vertical lines for division, similar in appearance to the spine of ancient books, dropping a hint of the relevance of the building's functions and properties.

The new building's 1F plane layout is similar to a windmill, where the " 回 " character shaped outer contour is divided into three parts with three main and secondary entrances respectively leading to the exhibition area, technical office area and auxiliary service area; 2F, 3F, 4F and 5F have a C-shaped plane layout, with the archives repositories, technical rooms and offices arranged horizontally, forming an enclosed space which is an accessible roof with unique landscape.

The new Archives' design provides both reasonable functions and novel appearance, which can adapt to the development of the archives business in the next 50 years of Lipu County.

1 档案馆外景 / The Archives Exterior

荔浦县档案馆

2 | 3
4 | 5
6

2 档案查阅室 / Archives Reference Room
3 入口大厅 / Entrance Lobby
4 目录室 / Catalogue Room
5 内庭院局部 / Inner Courtyard Local View
6 展厅 / Exhibition Hall

南宁市西乡塘区档案馆

Xixiangtang District Archives, Nanning

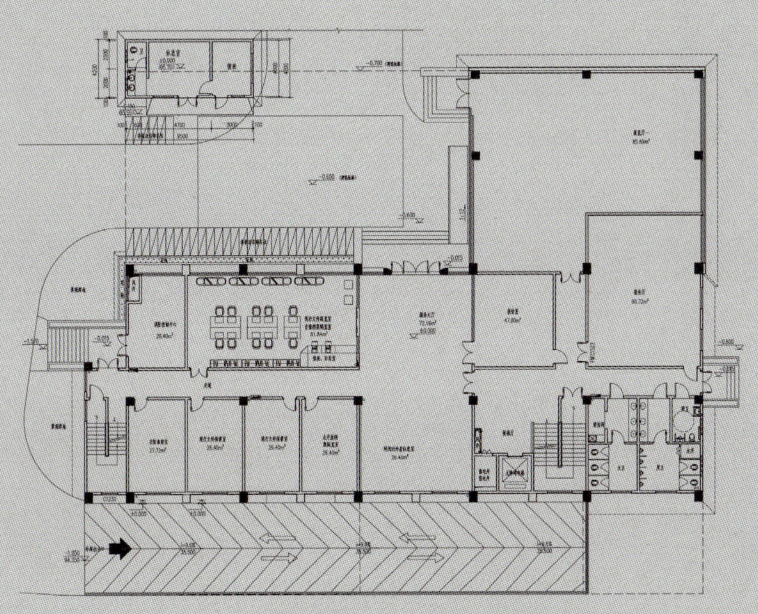

一层平面图 / Ground Floor Plan

占地面积：3333m²
建筑面积：5200m²
库房面积：1400m²
层　　数：主楼 5 层，地下 1 层
设计单位：广西华蓝设计（集团）有限公司设计
建成时间：2017 年 6 月

Site area: 3,333m²
Floor area: 5,200m²
Repository area: 1,400m²
Number of floors: ↑ 5 floors and ↓ 1 level
Time of completion: June 2017

南宁市西乡塘区档案馆成立于2005年8月，由原城北区档案馆和原永新区档案馆两馆合并而成，现有馆藏档案66个全宗，19000卷又355481件，档案起止年度为1966～2017年。

南宁市西乡塘区档案馆新馆位于南宁市西乡塘区兴津路与兴津一支路交叉口。档案馆新馆为单体建筑，外立面简洁、大方，兼顾广西民族特色的风格，并融入现代元素。南北立面凸出的横向与竖向分割墙，隐喻着“书”与“书架”的形象，暗示建筑性质，同时彰显区域性标志建筑的特征。

新馆按县级二类档案馆标准建设，矩形场地中，建筑平面呈L形布局，半围合出入口的缓冲空间。新馆总建筑面积5200m^2，其中，地下1层1450m^2，主要为停车场和变电用房；地上5层3750m^2。一层为对外服务用房，二层为档案业务和技术用房，三层、四层为档案库房，五层为办公用房。馆内配备防盗报警及智能门禁系统、高压细水雾自动灭火系统、温湿度监测系统等设施设备，保障了档案馆安全高效地运作，同时也体现了现代档案馆以人为本、服务为先的管理理念。

Xixiangtang District Archives of Nanning was established in August 2005 by incorporating the former Chengbei District Archives and former Yongxin District Archives. It has a collection of 66 fonds including 19,000 files 355,481 items, dating from year 1966 to 2017.

The new building of Xixiangtang District Archives is located at the intersection of Xingjin Road and the 1st Branch of Xingjin Road in Xixiangtang District, Nanning City. It is a single building with a simple, elegant facade combining national characteristics of Guangxi with modern elements. The protruding horizontal and vertical partition walls on the south and north facades are metaphorical for “books” and “bookshelves”, the essential functions of the building, while highlight the characteristics of the regional landmark building.

The building was constructed to the standards of a class-2 county-level archives. In the rectangular site, an L-shaped planar layout is adopted, with a semi-enclosed transitional space formed at the entrance. The building has a total building area of 5,200m^2, of which 1,450m^2 is underground, mainly used for the parking lot and substations; the five floors above ground count up to 3,750m^2, with 1F for public service rooms, 2F for archival business and technical rooms, 3F and 4F for archives repositories and 5F for offices. The building is equipped with anti-theft alarming and intelligent access control system, high-pressure water mist automatic fire-extinguishing system, temperature and humidity monitoring system, as well as and other facilities and equipment, ensuring the safe, efficient operation of the Archives and reflecting the people-oriented, service-oriented management concept of modern archives.

1 档案馆外景 / The Archives Exterior

市西乡塘区国家档案馆
南宁市西乡塘区档案局

2 | 3
4
5 | 6

2 展厅 / Exhibition Hall
3 档案查阅室 / Archives Reference Room
4 内庭院 / Inner Courtyard
5 入口 / Entrance
6 库房 / Repository

文昌市档案馆

Wenchang Archives

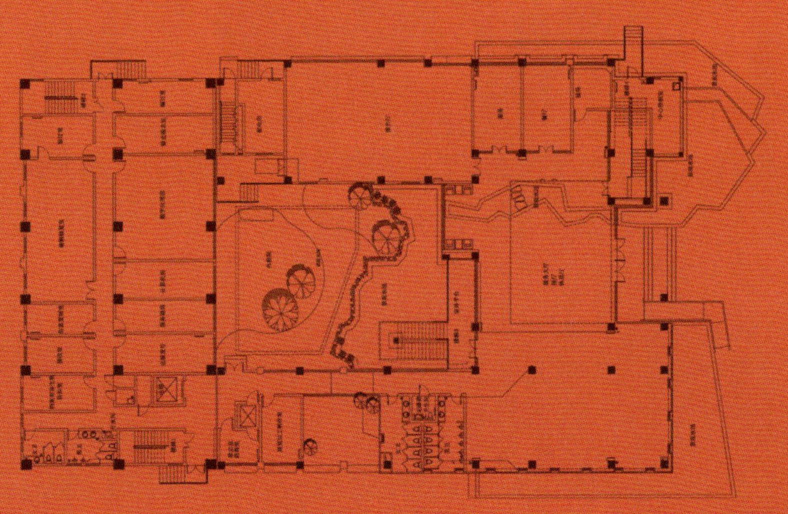

一层平面图 / Ground Floor Plan

占地面积：$6590m^2$
建筑面积：$3709m^2$
库房面积：$1002m^2$
层　　数：3层
设计单位：中船第九设计研究工程有限公司
建成时间：2015年1月

Site area: $6{,}590m^2$
Floor area: $3{,}709m^2$
Repository area: $1{,}002m^2$
Number of floors: 3 floors
Time of completion: January 2015

文昌市档案馆于1959年11月成立，馆藏档案126个全宗，文书档案18794卷又205684件，音像档案63盒、照片档案15册、各姓氏族谱87册、各类资料5000多册。

文昌市档案馆新馆位于文昌市文城镇文清大道西北侧，于2010年正式开工建设，2014年工程全部完工并通过验收，2015年正式搬迁启用。

档案馆新馆平面布局为“回”字形，内置庭院。建筑主立面以磨砂玻璃、镜面玻璃和实墙体块形成门厅灰空间，两种不同材质的玻璃将立面竖向分割，类似于“书脊”的形态，暗示档案馆这一建筑性质，同时磨砂玻璃的使用提供给室内充足采光，保证一定的隐私性；夹在磨砂玻璃体块之间的镜面玻璃为可开启扇，利于室内换气通风的同时，保障了建筑立面的统一与完整。

档案馆新馆为3层单体建筑，配有档案库房、业务和技术用房、对外服务用房、办公室和辅助用房。设计库房的容量为15万卷，基本可以满足30年的馆藏量，库房部分根据需要可以加至6层（设计中已考虑加层地基）。一层东侧有多功能会议室以及绿化、道路、室外停车场、围墙、水电配套设施等；南侧为对外服务区域：布有入口大厅、休息厅、展厅；北侧为档案馆技术用房区域：由档案接收、技术处理、档案整理和计算机房等组成；内部庭院为前后区域的过渡联系区域和消防景观两用水池。二层南侧为档案借阅区域：由借阅登记、目录室、开放档案借阅、未开放档案借阅、音像档案借阅、档案复印室等组成；北侧为档案库房区域。三层南侧为内部办公区域和部分功能用房；北侧为档案库房区域。

新馆值得一提的为内庭院的营造，门厅与内庭院以室外平台和景观水池相隔，形成了近、中、远景，增加景观层次，吸引访客对内庭院的探索欲望；于景观水池之上设置悬臂楼梯，巧妙将室外空间、半室外灰空间与室内空间融合在一起，起到过渡室内外的作用；院内营造的景观、种植的树木与三层上人平台形成景观视线通廊，连通、激活两个活动空间。

1 档案馆鸟瞰外景 / The Archives Aerial View

Established in November 1959, Wenchang Archives has a collection of 126 fonds, 18,794 files and 205,684 items of administrative archives, 63 audio-visual cassettes, 15 photo albums, 87 volumes of clan pedigrees, and over 5,000 volumes of all classifications of documentary archives.

The Archives' new building is located in the northwest side of Wenqing Avenue, Wencheng Town of Wenchang City. The construction was officially started in 2010, completed and delivered in 2014, and officially put into use in 2015.

The new building's planar layout is in the shape of Chinese character "回" with a courtyard within. On the main facade frosted glass, mirror glass and solid wall massing form a gray space at the hallway. The two different types of glasses divide the facade vertically to create a shape similar to "book spine", implying the architecture's nature. The frosted glass allows sufficient lighting indoors while ensures privacy; the mirror glass sandwiched between frosted glass massing is openable, which facilitates ventilation inside and maintain the unity and integrity of the facade.

The new Archives is a three-story single building comprising Archives repositories, Business and Technical Rooms, Public Service Rooms, Offices and Auxiliary Rooms. The repositories are designed to accommodate 150,000 files, which can basically meet the demand for 30 years. A sixth floor can be added to the building's repository part when necessary (which is considered in the foundation design). On the east side of 1F are the Multi-purpose Conference Room, as well as afforestation, roads, outdoor parking lot, enclosing wall, water and electricity facilities, etc.; in the south is the Public Services Area consisting of the Entrance Hall, Lounge, and Exhibition Hall; the north part is Technical Room Area comprising the Archives Accession Room, Technical Processing Room, Archives Sorting Room and Computer Room; the courtyard inside, where a fire-landscape water pool is built, links the front zone with the rear zone. The south part of 2F is Archives Borrowing Area, consisting of the Borrowing Registration Room, Cataloging Room, Open Archives Borrowing, Unopened Archives Borrowing, Audio-visual Archives Borrowing, and Archives Photocopying Room; archives repositories are at the north side. On the south side of 3F are the Internal Office Area and some functional rooms; in the north are archives repositories.

The new building's inner courtyard is quite a spotlight. An outdoor terrace and the landscape water pool is arranged between the courtyard and the landscape water pool, forming a landscape with rich layers from near view, medium view to distant view, attracting the visitors to explore the courtyard inside; a cantilever staircase is ingeniously built above the landscape pool, acting as a transition that mixes together the outdoor spaces, semi-outdoor gray space and indoor spaces; landscape in the courtyard, trees planted and the walkable platform on 3F link up to become a landscape sight corridor, connecting and activating the two activity spaces.

2 | 3
4
5 | 6

2 内院 / Inner Courtyard
3 入口大厅 / Entrance Lobby
4 展厅 1 / Exhibition Hall 1
5 展厅 2 / Exhibition Hall 2
6 库房 / Repository

涪陵区档案馆

Fuling District Archives

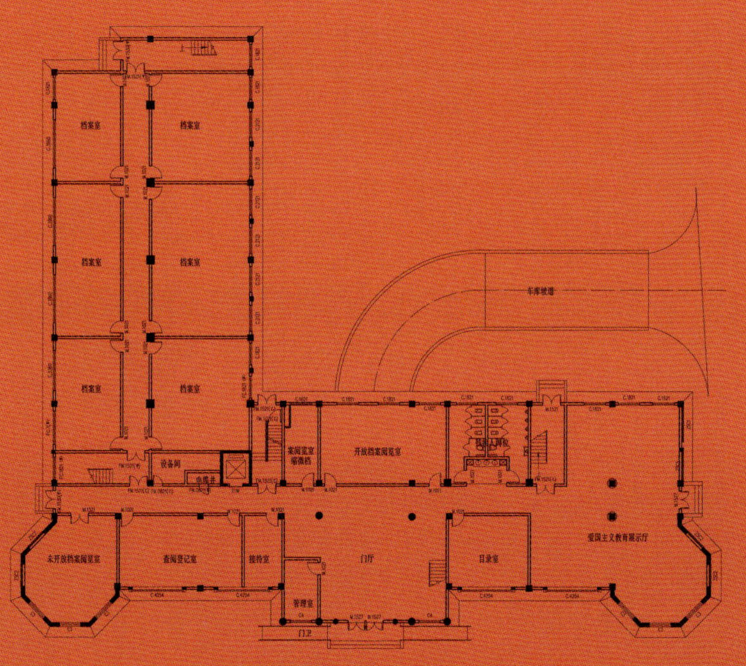

一层平面图 / Ground Floor Plan

占地面积： 6660m²
建筑面积： 7378m²
库房面积： 3000m²
层　　数： 主楼 5 层，地下 1 层
设计单位： 重庆市工程设计院
建成时间： 2012 年 9 月

Site area: 6,660m²
Floor area: 7,378m²
Repository area: 3,000m²
Number of floors: ↑ 5 floors and ↓ 1 level
Time of completion: September 2012

涪陵区档案馆，现有馆藏档案全宗599个，档案902110卷（册）。2013年12月成功创建国家一级综合档案馆。

涪陵区档案馆新馆独立建设，位于涪陵新城区东西主干道交汇处。于2008年立项，2012年8月竣工验收，同年9月投入使用。

档案馆新馆按照涪陵新城区“保山、藏水、控高、扮靓”的建设原则总体规划设计。建筑呈现L形布局，东南侧体块为档案业务技术办公区，东北侧体块为档案库房区，半围合的内院依山而建，在保证基地原有的自然环境风貌真实性的基础上，营造环境优美的院落空间。主体建筑外观仿哥特式建筑欧式风格，竖向线条分割明显，整体庄重、统一，形成向上的动势。附属建筑同主体建筑、周边环境相互衬托、互为补充，充分展现时代风貌、地域特点、现代水准的整体效应。

新馆一层为档案查阅、展厅等对外服务功能，二～五层主楼主要为技术办公和行政办公用房，配楼为档案库房，便于档案管理，利于技术办公区便捷地到达库房区，提高档案馆的运行工作效率。

Fuling District Archives has a collection of 599 fonds and including 902,110 files (volumes). In December 2013, it is upgraded to a National Class-1 Archives.

Fuling District Archives' new building is located at the intersection of the east and west trunk roads of the new urban areas of Fuling. The project was approved in 2008, completed in August 2012 and put into use in September in the same year.

The new building is planned and designed according to the general construction principles of the new urban area of Fuling, which is "Preserve the hills, conceal the waters, and control the height to make the city more beautiful". The building takes an L-shaped layout, with its southeast massing being the archival business technical office area, and the northeast massing being the archives repository area; the semi-enclosed inner courtyard is built leaning the mountain, creating a beautiful space where the original natural environment style and features are perfectly maintained. The main building takes a European Gothic-style appearance, with apparent vertical dividing lines used to present an upward momentum in the overall solemnity and unity. The ancillary buildings, main building and ambient environment set off and complement each other, fully reflecting the overall design effects that incorporate the characteristics of the times, geographical features and modern standards.

1F of the new building is for public services like archives reference, Exhibition Halls, etc. 2F to 5F are mainly for Technical Offices and Administrative Offices. The wing building accommodates Archives Repositories, which enables easy management and convenient access from the technical offices, improving operation efficiency of the Archives.

1 档案馆外景 / The Archives Exterior

2 | 3
4
5

2 入口大厅 / Entrance Lobby
3 展厅/ Exhibition Hall 1
4 档案查阅大厅 / Archives Reference Hall
5 展厅 2 / Exhibition Hall 2

合川区档案馆

Hechuan District Archives

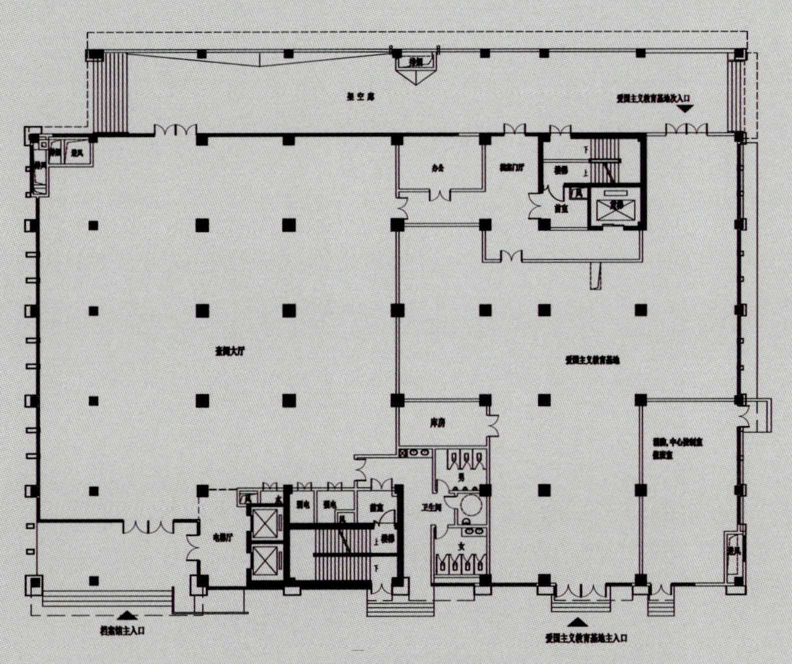

一层平面图 / Ground Floor Plan

占地面积： 6352m²
建筑面积： 9632m²
库房面积： 4000m²
层　　数： 主楼15层，配楼2层，地下1层
设计单位： 重庆大学建筑设计研究院
建成时间： 2016年9月

Site area: 6,352m²
Floor area: 9,632m²
Repository area: 4,000m²
Number of floors: 15 floors for the main building, 2 floors for the wing and ↓ 1 level
Time of completion: September 2016

合川区档案馆成立于1959年3月，2007年命名为合川区爱国主义教育基地，2012年获得国家二级综合档案馆称号。现馆藏档案31万余卷（件），资料46730卷（册）。

合川区档案馆新馆为独栋高层建筑，位于合川区新城区国土局大院，2014年7月动工，2016年9月竣工，2017年7月投入使用。建筑塔楼采用石材与玻璃结合的形式，立面使用竖向条窗，增强塔楼竖向线条，使得塔楼视觉上更为挺拔、高耸。

功能布局上，裙房中主要布置查阅大厅、主题展览等较为开放的功能空间；塔楼主要为行政办公、技术办公和库房功能。塔楼将核心筒安放于两个端头，库房形态保持完整；周边有走廊环绕，少外墙，通过走廊的缓冲空间分隔，避免室外环境对库房的干扰。设计理念以减少占地、增加空间为重点；风格特色以绿色生态、节能环保为出发点，营造舒适宜人的建筑外部空间。

The original Hechuan District Archives of chongqing was established in March 1959. In 2007, it was awarded Hechuan District Patriotism Education Base, and in 2012 National Class-2 General Archives. It has a collection of more than 310,000 files (items) with 46,730 volumes (volumes) of documentary archives.

The new site of Hechuan District Archives is an independent high-rise building construction project, located in the compound of the Land and Resources Bureau of Hechuan's new urban area. The project was commenced in July 2014, completed in September 2016, put into use in July 2017. The tower building, combining stone and glasses, has a facade with vertical long windows on it, making the building looks more lofty and towering.

In terms of functional layout, the podium mainly provide relatively open functions, such as the Reference Hall and subject exhibitions; the tower building is chiefly for Administrative Offices, Technical Offices and repositories. The tower building, with its core tube arranged at both ends, keeps all repositories in an integral part, which are surrounded by corridors acting as separation and transition, thus reducing the number of outer walls and avoid interference from the outdoor environment to the repositories. The design, focusing on reduced land occupation and increased spaces, aims at a style that is green, ecological, energy conserving and environmental protective, so as to create a attractive, comfortable exterior space for the building.

1 档案馆外景 / The Archives Exterior

2 档案馆入口 / The Archives Entrance

3
4 | 5
6 | 7

3 展厅 / Exhibition Hall
4 档案查阅大厅 / Archives Reference Hall
5 中控室 / Central Control Room
6 库房 / Repository
7 档案查阅室 / Archives Reference Room

成都市郫都区档案馆

Pidu District Archives, Chengdu

Sichuan Province

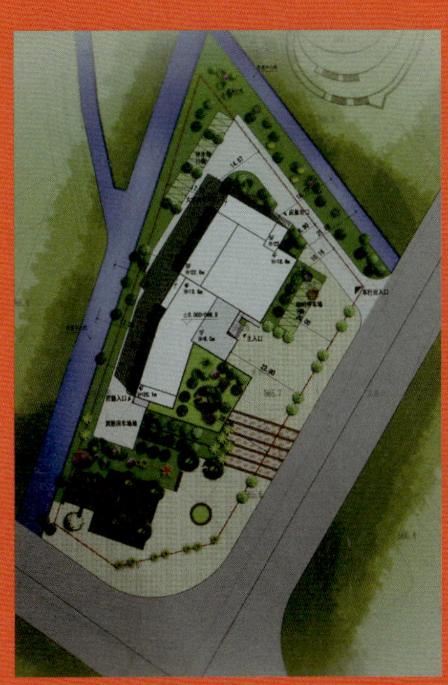

总平面图 / Master Plan

占地面积：6047m²
建筑面积：3991m²
库房面积：2300m²
层　　数：5 层
设计单位：汉嘉设计集团股份有限公司
建成时间：2012 年 9 月

Site area: 6,047m²
Floor area: 3,991m²
Repository area: 2,300m²
Number of floors: 5 floors
Time of completion: September 2012

郫都國家綜合檔案館
成都市郫都
535

1 档案馆外景 / The Archives Exterior

成都市郫都区档案馆成立于1959年，馆藏档案11万卷、65万件、图书资料5400余册，2014年晋升为国家一级综合档案馆，是郫都区爱国主义教育基地。

成都市郫都区档案馆新馆为单体建筑，坐落于郫都区德源北路二段535号。建筑以砖红色为主要色彩，两个方形体块交界处设置主入口，主立面底部营造柱廊灰空间，配合竖条形开窗，增强建筑竖向分割，树立挺拔的形象。

新馆功能较为完善，一层设有“蜀源郫都历史文化档案陈列展”“廉文化档案陈列展”两个展厅以及档案查询服务大厅等，以较为开敞的空间和较为开放的功能展现新馆积极开放的状态；二、三、四层设置档案库房、技术办公及行政办公等。功能布局动静分区合理，保证了档案馆各部门高效合理地运作。

Pidu District Archives of Chengdu was established in 1959. The Archives has a collection of 110,000 files, 650,000 items and over 5,400 books. Upgraded to a national class-1 general archives in 2014, the Archives is the patriotic education base of Pudu District.

The new building of Pidu District Archives is a single building seated at No. 535, Section 2 of Deyuan North Road, Pidu District. Brick red is used as the building's main color. The main entrance is arranged at the junction of two square masses, and a colonnade gray space is formed at the bottom of the main facade. Vertical strip windows on the facade enhance the vertical partitioning of the building and endows it with a lofty, eminent image.

The new building provides relatively complete functions. On 1F there are two exhibition halls, namely “Historical and Cultural Archives Exhibition of Pidu as the Origin of the Shu Kingdom “and” Incorruption Culture Archives Exhibition”, as well as the Archives Reference and Service Hall, showcasing the new Archives' positiveness and openness with relatively open spaces and functions. 2F, 3F and 4F accommodate the Archives Repositories, Technical Offices and Administrative Offices. The functional layout is reasonably designed to separate dynamic zones from static ones, ensuring efficient and reasonable operation of all departments of the Archives.

2 | 3
4
5 | 6

2 入口 / Entrance
3 文化走廊 / Culture Gallery
4 展厅 1 / Exhibition Hall 1
5 展厅 2 / Exhibition Hall 2
6 库房 / Repository

德昌县档案馆

Dechang Archives

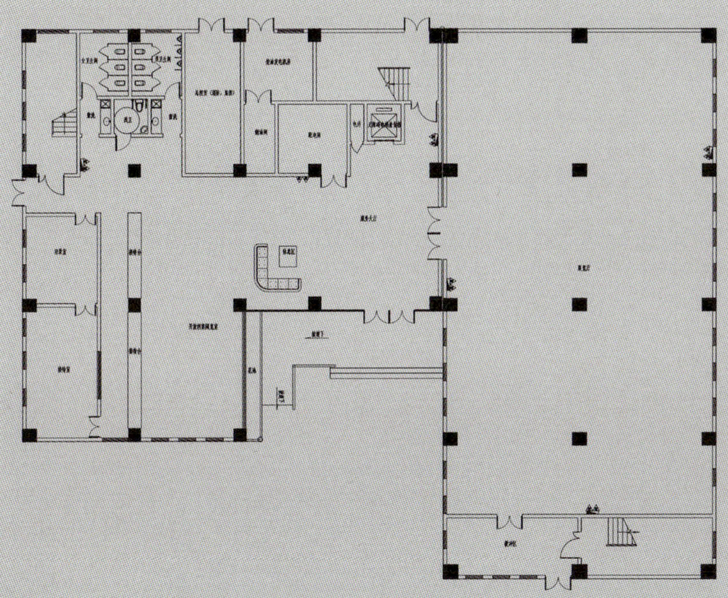

一层平面图 / Ground Floor Plan

占地面积：4230m²
建筑面积：4583m²
库房面积：1746m²
层　　数：主楼 5 层，配楼 4 层
设计单位：四川国信建筑设计有限公司
建成时间：2017 年 3 月

Site area: 4,230m²
Floor area: 4,583m²
Repository area: 1,746m²
Number of floors: 5 floors for the main building and 4 floors for the wing
Time of completion: March 2017

德

德昌县档案馆成立于1963年12月，现有馆藏档案108个全宗、85711卷（册）、47458件，照片档案4511张，录音录像档案341盘，实物档案994件，资料16597册。

德昌县档案馆新馆为独立建筑，位于德昌县德州镇南街208号。建筑总平面采用C形布局，两侧为墙体较多的“实体块”，中间夹以玻璃“虚体块”，凹入部分形成建筑主入口，以黄色实墙和墨绿色玻璃为主，整体统一，简洁大方。

新馆布局合理，功能齐全，一层为展览、服务大厅、开放档案阅览室等较为开敞的公共空间；二层及以上楼层，主楼为库房及特藏室，配楼为办公区、技术用房及对外服务用房，上下动静分离，能够满足档案馆功能需求，保证档案馆合理高效运作。

Established in December 1963, Dechang Archives has a collection of 108 fonds, 85,711 files (volumes) and 47,458 items, as well as 4,511 photos, 341 audio and visual disks, 994 material objects and 16,597 volumes of documentary archives.

The new building of Dechang Archives is an independent building located at No. 208 Zhennan Street, Dezhou Town of Dechang County. It adopts a C-shaped overall planar layout, with solid blocks with more walls on both sides, and glass "virtual blocks" in the middle. The concave part forms the main entrance of building, whose entirety is dominated by yellow solid walls and dark green glasses in a unified, simple and elegant style.

The new Archives has a reasonable layout and complete functions. 1F consists of relatively open public spaces such as the Exhibition and Service Hall, Public Archives Reading Room, etc. On 2F and above floors, the part on the main building accommodates repositories and special collection rooms, and the part on the wing building is for office area, technical rooms and public service rooms, with dynamic areas separated from static ones to meet the functional requirements of the Archives and ensure its operation in a reasonable, efficient manner.

1 档案馆外景 / The Archives Exterior

2
3 | 4
5 | 6

2 接待大厅 / Reception Hall
3 档案消毒杀虫室 / Archives Disinfection Room
4 入口 / Entrance
5 档案查阅大厅 / Archives Reference Hall
6 库房 / Repository

富顺县档案馆

Fushun Archives

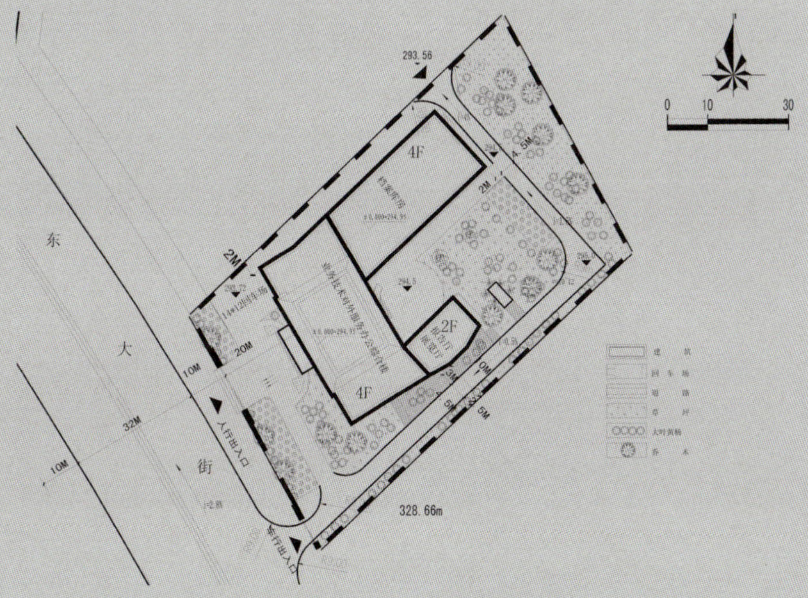

总平面图 / Master Plan

占地面积：10005m^2
建筑面积：6800m^2
库房面积：2800m^2
层　　数：主楼 4 层，配楼 2 层
设计单位：四川远建建筑工程设计有限公司
建成时间：2017 年 12 月

Site area: 10,005m^2
Floor area: 6,800m^2
Repository area: 2,800m^2
Number of floors: 4 floors for the main building and 2 floors for the wing
Time of completion: December 2017

富顺县档案馆设立于1959年4月，现有馆藏档案372个全宗、17万余卷、16万余件，资料1万余册。

富顺县档案馆新馆为独立单体建筑，位于富顺县东湖镇东大街。建筑总平面采用U形布局，整体建筑设计典雅庄重，三边半围合出中心广场与景观；立面造型以门厅为轴线对称，形成两侧突出，中间凹入的竖向三段层次；门厅入口挑出于立面，突出主入口的显著地位，增强建筑仪式感；立面两侧的浮雕，体现了档案馆的历史文化元素，使档案馆成为当地标志性文化建筑。

新馆功能布局合理，一层为展览、服务大厅、开放档案阅览室等较为开敞的公共空间，门厅设计端庄大气，流线合理，二至四层为库房、技术办公和行政办公区，动静分区合理，保证档案馆合理高效地运作。

Established in April 1959, Fushun Archives has 372 fonds, more than 170,000 files and over 160,000 items of archives, as well as over 10,000 volumes of documentary archives in its holding.

The new building of Fushun Archives is a single building located in the East Street, Donghu Town of Fushun County. The building adopts a U-shaped general planar layout. In an overall elegant, solemn architectural design, the building's three sides forms a semi-enclosure with the central square and landscape inside; the facade is axisymmetric around the entrance hall, forming a vertical three-part architectural structure that is protruding at both sides and concave in the middle. The hallway entrance sticks out of the facade, highlighting the prominence of the main entrance and creating a sense of ceremony; reliefs on both sides of the facade reflect the Archives' historical and cultural elements, making it a local cultural landmark.

The new building has a reasonable functional layout, with 1F accommodating the relatively open public spaces such as the Exhibition and Service Hall, Public Archives Reading Room, etc. The hallway has a dignified, grand design with reasonably arranged streamlines. 2F through 4F accommodate the repositories, technical offices and administrative offices, with dynamic areas separated from static ones to the Archives' reasonable, efficient operation.

1 档案馆鸟瞰 / The Archives Aerial View

2 档案馆入口 / The Archives Entrance

3
4 | 5
6

3 展厅 1 / Exhibition Hall 1
4 展厅 2 / Exhibition Hall 2
5 档案查阅大厅 / Archives Reference Hall
6 库房 / Repository

米易县档案馆

Miyi Archives

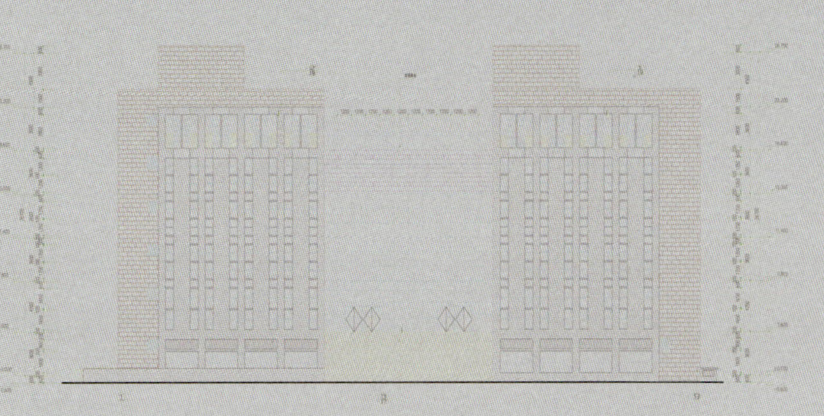

正立面图 / Front Elevation Plan

占地面积：3335m^2
建筑面积：6650m^2
库房面积：1917m^2
层　　数：主楼 6 层，地下 1 层
设计单位：上海同济大学
建成时间：2012 年 5 月

Site area: 3,335m^2
Floor area: 6,650m^2
Repository area: 1,917m^2
Number of floors: ↑ 6 floors and ↓ 1 level
Time of completion: May 2012

米易县档案馆成立于 1959 年 11 月，是国家一级档案馆、全国中小学档案教育社会实践基地。

米易县档案馆为单体建筑，坐落在县城新区安宁路 76 号。建筑总平面采用扇形布局，两个实体块中间为虚体块相交，中间玻璃体块作为门厅入口，并结合虚体块的优势将门厅做成通高空间，由室外拾阶而上进入门厅，形成了多层次的空间变化关系。建筑立面开窗注重竖向分隔，将实墙自上而下连续，增强建筑向上的动势，渲染档案馆高大、挺拔的形象，是当地文化标志性建筑。

新馆设计理念新颖，功能完备，入口空间布置展览、服务大厅等较为开敞的公共空间，其余楼层布置档案库房、技术用房、行政用房等，动静分离，布局合理。建筑技术上，运用了先进的建筑橡胶隔震支座技术，配备了多联机中央空调、七氟丙烷、全域监控、温湿度监控、门禁、无轨密集架等设施设备，保证档案馆安全、合理、高效地运行。

1 档案馆外景 / The Archives Exterior

Established in November 1959, Miyi Archives is national class-1 archives as well as national social practice base for archives education in primary and secondary schools.

The Archives is a single building located in No. 76 Anning Road, the new town area of Miyi County. The building has a fan-shaped general layout, with intersecting virtual masses between two physical masses, and glass massing in the middle used as the hallway entrance. Taking advantages of the virtual massing, the hallway is designed as a well-hole style space, which is accessed from outdoor steps, forming multi-layered spatial variations. The facade features vertical separation and is made of continuous solid walls from top to bottom, which enhances the upward momentum of the building, renders a towering, lofty image of the Archives as one of local cultural landmarks.

The new building is designed with novel concepts and full-scale functions. The entrance space serves for exhibitions, Service Hall and other open public services, and the other floors are used for Archives Repositories, Technical rooms and Offices, etc. to combine dynamics with stillness and offer a reasonable layout. It also utilizes advanced construction technologies such as the rubber isolation bearing for construction, and equipped with VRF central air conditioning, heptafluoropropane fire extinguishing system, global monitoring, temperature and humidity monitoring, access control, trackless compact shelving and other facilities to ensure safe, reasonable and efficient operation of the Archives.

2 | 3
4
5 | 6

2 档案查阅大厅 / Archives Reference Hall
3 多功能厅 / Multi-purpose Hall
4 展厅 1 / Exhibition Hall 1
5 资料室 / Documentary Archives Room
6 展厅 2 / Exhibition Hall 2

瓮安县档案馆

Weng'an Archives

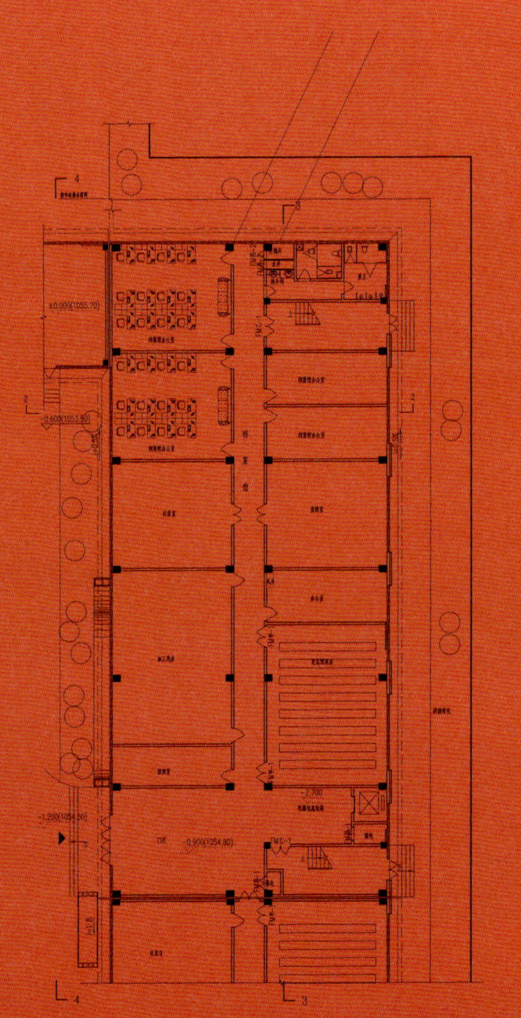
一层平面图 / Ground Floor Plan

占地面积：968m^2
建筑面积：2850m^2
库房面积：964m^2
层　　数：5 层
设计单位：湖南省湘潭市建筑设计院
建成时间：2017 年 8 月

Site area: 968m^2
Floor area: 2,850m^2
Repository area: 964m^2
Number of floors: 5 floors
Time of completion: August 2017

瓮安县档案馆 1959 年 5 月建立，现内设综合科、业务法规科、接收管理利用科，下设瓮安县现行文件资料服务中心，馆藏档案 131 个全宗，约 5 万卷（件）。

瓮安县档案馆新馆坐落于瓮安县雍阳街道办事处城北社区，与方志馆、图书馆和新闻中心整合修建。

档案馆新馆设计风格现代简约，采用铝板横向层层分割，如同堆积的档案袋，又如堆叠的书籍，立面造型对档案馆及图书馆的建筑性质起到一定暗示作用。整体建筑风格简洁明快，注重时代特征，与周边环境及建筑风格相统一，是一座现代感较强的档案馆。

档案馆功能分区明确合理，一层主要设置展厅、查阅室等对外服务功能，二层为库房，三层为办公及库房等，功能布局合理。

新馆坐落区域为全县统一规划的文体活动中心，档案馆、图书馆、方志馆整体设计，营造出温馨和谐的办公环境。

Weng'an Archives was established in May 1959, currently consisting of the Comprehensive Department, Business Laws and Regulations Department, as well as the Accession, Management and Utilization Department, with the subordinate Prevailing Documentation Service Center of Weng'an County. The Archives has 131 fonds, with a total collection of about 50,000 files (items).

The new building of Weng'an Archives is located in the Chengbei Community of the Yongyang Sub-district Office in Weng'an County, constructed as part of the Project which also comprises the Local Chronicles Museum, Library and the Press Center.

The new Archives, adopting a modern, simple design style, is divided into horizontal layers by aluminum sheets, creating an appearance like stacked archives portfolios or piled-up books. The facade in a certain degree implies the architectural nature of the Archives and Library. The overall architectural style, being simple and clear, focuses on the characteristics of the times and is consistent with the surrounding environment and architectures, endowing the building with a strong modern sense.

The functional division of the Archives is clear and reasonable. 1F mainly provides public service functions such as the Exhibition Hall and Reference Room. 2F accommodates repositories, and 3F is for offices and repositories, forming a quite reasonable functional layout.

The zone is planned as the county's cultural and sports center, with an overall design applied to the Archives, Library and Local Chronicles Museum to create warm, harmonious office environment.

1 档案馆鸟瞰外景 / The Archives Aerial View

追 赶 领 先
冠珠陶瓷
大自然地板
TATA木门

2 | 3
4
5

2 库房 / Repository
3 一层大厅 / GF Lobby
4 档案查阅室 / Archives Reference Room
5 会议室 / Conference Room

遵义市汇川区档案馆

Huichuan District Archives, Zunyi

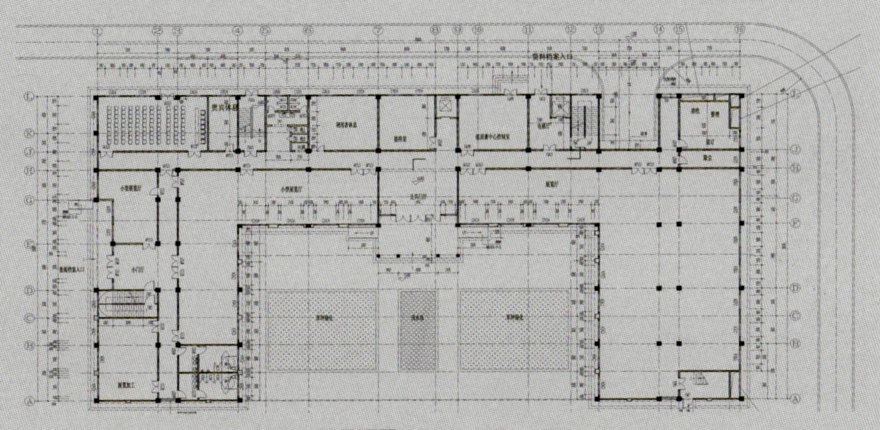

一层平面图 / Ground Floor Plan

旧馆 Old

New 新馆

占地面积： $10000m^2$
建筑面积： $10857m^2$
库房面积： $3628m^2$
层　　数： 7 层
设计单位： 遵义市建筑规划设计院有限责任公司
建成时间： 2017 年 2 月

Site area: $10,000m^2$
Floor area: $10,857m^2$
Repository area: $3,628m^2$
Number of floors: 7 floors
Time of completion: February 2017

遵义市汇川区档案馆新馆是档案馆和方志馆合二为一的综合性建筑，位于与汇川区交通主干道汇川大道相交汇的红河北路中段，紧邻汇川区行政办公中心，正对风景优美的莲花山森林公园。

遵义市汇川区档案馆新馆本着安全、长久、经济适用、美观的设计原则，力争 30 年不落后的高标准建设要求进行设计。建筑图底关系上，采用“凹”字形总平面布局，两侧配楼与主楼围合出中央景观，花园采用假山和水体喷泉的园林式组合，栽种了香樟、银杏等上百种代表四季的花草树木，将档案馆置于一个优美、干净的环境中，与莲花山森林公园形成对应关系，同时也通过对称式花园的营建，增强建筑氛围的庄重感；建筑造型上，充分考虑建筑与周边环境的协调，设计上以灰色为主基调，体现民国风与黔北地方建筑元素手法。

新馆按规划标准设置了档案库房、档案消毒除尘室、整理室、档案查阅大厅等 30 余个功能性用房。负一层是车库，一层是展厅和功能用房，二层是库房和业务用房，三、四、五层是档案库房，六层是业务用房。新馆制定了使用年限为 50 年，建筑等级为三级的施工方案，保证了档案馆安全高效地运作。

新馆的建成极大地提升了汇川区档案馆的馆藏容量和功能作用，较大地改善了档案安全硬件保障条件。

2 档案馆外环境 / The Archives Outdoor Environment

The new Archives of Huichuan District Archives is a complex project which also comprises the Local Chronicles Museum, seated in the middle section of Honghe North Road, at the intersection with Huichuan Avenue, the traffic trunk of Huichuan District. It is adjacent to the Administrative Office Center of Huichuan District, and faces the scenic Lianhua Hill Forest Park.

The building was designed with the principles of "safety, longevity, affordability and aestheticism" and constructed in accordance with high-standard requirements to design and to keep it capable of adapting the needs for the next 30 years. In terms of figure-ground relationship, a concave shaped general layout is adopted. Enclosed by the wing buildings at both sides and the main building a central garden is formed. In the garden rockeries, water and fountains constitute the landscape, with over one hundred species of flower and tree such as camphor and ginkgo planted, creating an exquisite, neat environment echoing the Lianhua Hill Forest Park. The symmetrical garden also enhances the building's solemnness. To harmonize the architectural style with the surrounding environment, a gray tone is adopted, reflecting the architectural characteristics of the Republic of China and features of the northern Guizhou area.

According to the planning standards, more than 30 functional rooms were designed for the new Archives, including Archives Repositories, Archives Disinfection and Dust Removal Room, Sorting Room, Archives Reference Hall, etc. Basement floor 1 is the garage, 1F is the Exhibition Hall and functional rooms, 2F accommodates repositories and business rooms, 3F, 4F and 5F are archives repositories, and 6F is for business rooms. The new Archives' construction plan was based on a service life of 50 years and building level 3, ensuring the safe and efficient operation of the Archives.

The completed new Archives offers greatly increased holding capacity and functions of Huichuan District, with substantially improved collection security conditions.

1 档案馆外景 / The Archives Exterior

3
4 | 5
6 | 7

3 会议室 / Conference Room
4 库房 / Repository
5 一层大厅 / GF Lobby
6 展厅 / Exhibition Hall
7 档案查阅大厅 / Archives Reference Hall

2
3 | 4
5 | 6

2 一层大厅 / GF Lobby
3 入口 / Entrance
4 档案查阅室 / Archives Reference Room
5 报告厅 / Lecture Hall
6 中控室 / Central Control Room

沧源佤族自治县档案馆

Archives of Cangyuan Wa Autonomous County

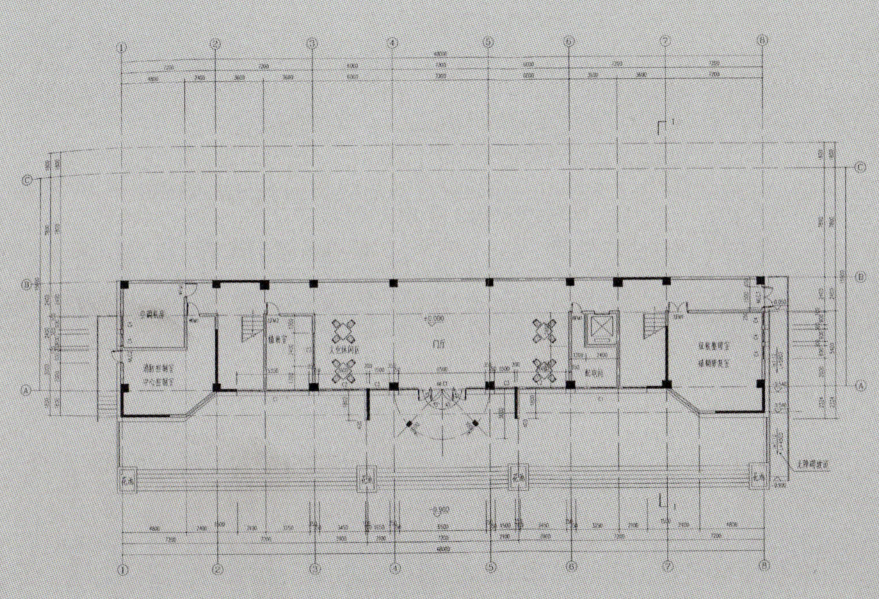

一层平面图 / Ground Floor Plan

占地面积：4725m²
建筑面积：3675m²
库房面积：1785m²
层　　数：5 层
设计单位：云南博超建筑设计有限公司
建成时间：2016 年 6 月

Site area: 4,725m²
Floor area: 3,675m²
Repository area: 1,785m²
Number of floors: 5 floors
Time of completion: June 2016

沧源佤族自治县档案馆现有馆藏档案 130 个全宗，63487 卷（盒、册、枚）814520 件，拥有档案目录数据 238 万条，档案数据存储量 19T。

沧源佤族自治县档案馆新馆位于沧源县勐董镇，临近沧源县县委大楼，建筑整体顺应地势，内部设有对外服务用房、档案业务和技术用房、办公室用房等，其中门厅为一、二层的通高空间，营造了便于交流、休憩的公共氛围；二层主要为展厅和阅览等对外功能区；三～五层相似，采用内廊式布置库房和内部办公用房。建筑功能采用竖向分区，减少干扰并且使用方便。

新馆的入口元素来源于具有 3500 多年历史的国家级重点文物——沧源崖画，延续了佤族文化特色。建筑主体采用浅黄色石材，彰显了公共文化建筑的庄重大气。建筑立面具有韵律感的竖条窗契合了档案馆的功能需求。新馆设计通过构图、材料、质感及色彩等各个方面相配合，强化了其城市地位。

1 档案馆外景 / The Archives Exterior

Archives of Cangyuan Wa Autonomous County has 130 fonds, with a total collection of 63,487 archival files (disks, volumes, pieces) and 814,520 archival items, 2,380,000 pieces of archival catalog data and an archival data storage of 19T.

The new building of Cangyuan Archives is located in Yudong Town of Cangyuan County, adjacent to the County Party Committee's Building. The entire building is built to adapt to the terrain conditions, with public service rooms, archival business and technical rooms, offices, etc. provided inside. The hallway is a full-height space occupying 1F and 2F, creating an open atmosphere for easy communication and rest; 2F mainly provides public functions such as Exhibition Hall and reading area; similarly in arrangement, 3F to 5F accommodate inner-corridor layout repositories and internal office rooms. The building features vertical function partitioning for minimum interference and ease of use.

The entrance elements of the new building come from Cangyuan cliff painting, one of the national key cultural relics with a history of more than 3,500 years; as an extension of the characteristics of the Wa nationalities culture, light yellow stones are used for the main body of the building, highlighting the grandeur of a public cultural architecture; the rhythmical vertical long windows on the building's facade agree with the functional requirements of the archives. The combination of the picture composition, materials, textures and colors in the new building's design strengthens its significance in the city.

2 | 3
4 | 5
6 | 7

2 展厅 1 / Exhibition Hall 1
3 库房 / Repository
4 展厅 2 / Exhibition Hall 2
5 入口 / Entrance
6 报告厅 / Lecture Hall
7 档案查阅室 / Archives Reference Room

昌宁县档案馆

Changning Archives

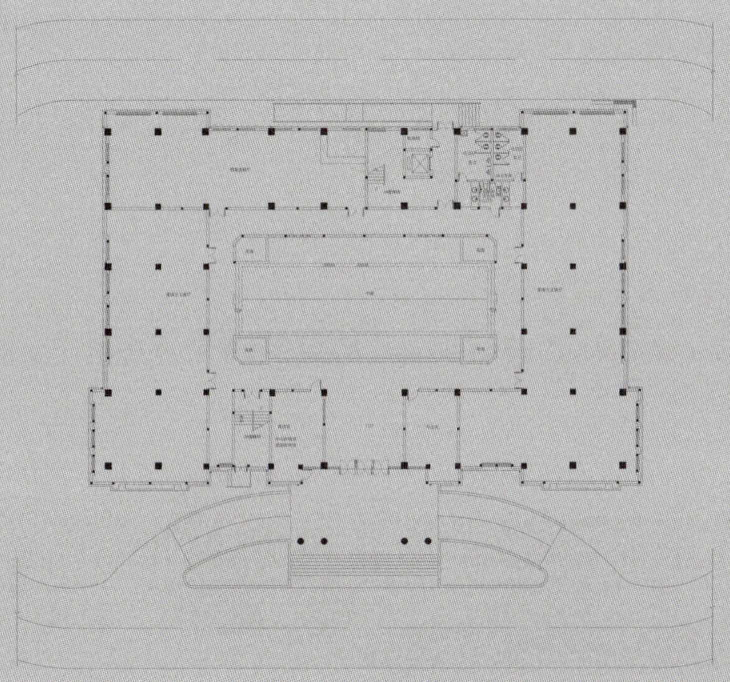

一层平面图 / Ground Floor Plan

占地面积：6666m²
建筑面积：5017m²
库房面积：1407m²
层　　数：4 层
设计单位：云南省轻纺工业设计院
建成时间：2015 年 12 月

Site area: 6,666m²
Floor area: 5,017m²
Repository area: 1,407m²
Number of floors: 4 floors
Time of completion: December 2015

昌宁县档案馆始建于1959年9月，现有馆藏文书档案159个全宗80584卷303286件，声像档案890盘7306张，实物档案1230枚244件，资料8362册。

昌宁县档案馆新馆位于田园镇滨河东路，毗邻体育馆，与文化馆和图书馆相互联系，是昌宁文化地标建筑之一。新馆为4层框架结构，按照县级一类馆标准进行设计，设计使用年限为50年。项目设计以构建“数字化、园林化、现代化”的一流档案馆为基本理念。建筑平面采用“口”字形分散式布局，具有明确的向心性，一层主要为对外服务用房，包括序厅、历史记忆厅、爱国情怀厅、多彩民族厅、千年茶乡厅、田园昌宁厅和报告厅等，二~四层为内部功能用房，主要为技术用房、办公用房以及库房等。

建筑采用中国传统的坡屋顶，错落有致，赋予了新馆新中式的建筑风格，立面采用双柱式，强化了入口形象。整体造型大气凝练、端庄典雅，具有深厚凝重的文化感和历史感，成为展现地域文化特色的精彩窗口，为昌宁县城市文化建设增添一抹亮色。

1 档案馆外景 / The Archives Exterior

Established in September 1959, Changning Archives has a collection of administrative archives including 159 administrative fonds, 80,584 files and 303,286 items, as well as 890 audio and visual disks and 7,306 pieces, 1,230 pieces and 244 items of material objects, as well as 8,362 volumes of documentary archives.

The new building of Changning Archives is located in East Binhe Road of Tianyuan town, adjacent to the gymnasium and linked with the Cultural Center and Library. It is one of the cultural landmarks of the county. Being a four-story framework structure, the new building is designed up to the standards of county-level Class 1 archives with a service life of 50 years. The project design is made following the fundamental philosophy to built it into a "digital, garden and modern" first-class archives. The building adopts a decentralized square planar layout with clear centripetality. 1F is mainly for public service rooms, including the Preface Hall, Historical Memory Hall, Patriotism Hall, Colorful Ethnic Hall, Millennium Tea Town Hall, Garden Changning Hall, Lecture Hall, etc.; 2F to 4F accommodate internal functional rooms, mainly the technical rooms, offices and repositories.

The building adopts the traditional Chinese pitched roof, which is in picturesque disorder and endows it with a new Chinese style. The double-column facade strengthens the image of entrance. The overall shape of the Archives is grandeur, concise and elegant with profound cultural and historical connotations, making it a wonderful window to showcase the regional cultural characteristics, and, adds luster to the urban culture construction of Changning County.

2
3
4 | 5

2 一层大厅 / GF Hall
3 展厅 / Exhibition Hall
4 库房 / Repository
5 档案查阅室 / Archives Reference Room

楚雄市档案馆

Chuxiong Archives

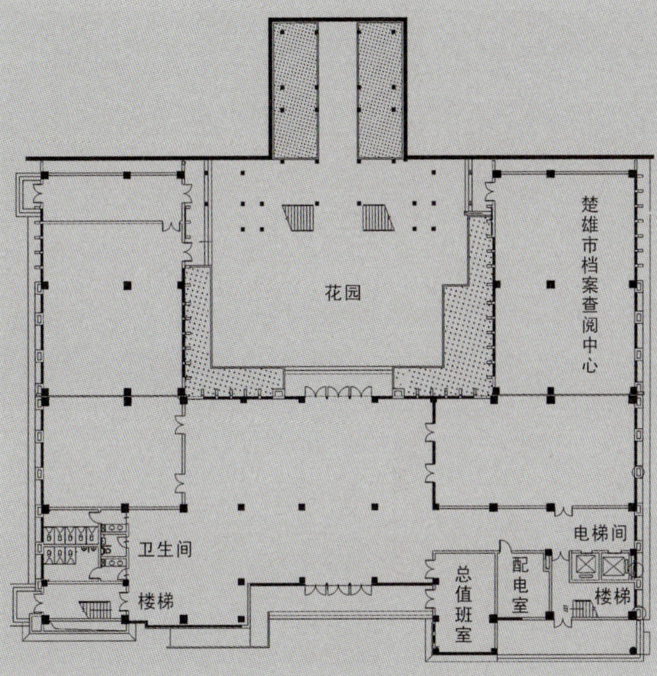

一层平面图 / Ground Floor Plan

占地面积：11774m^2
建筑面积：11068m^2
库房面积：5574m^2
层　　数：6 层
设计单位：广州景森设计公司
建成时间：2012 年 10 月

Site area: 11,774m^2
Floor area: 11,068m^2
Repository area: 5,574m^2
Number of floors: 6 floors
Time of completion: October 2012

楚雄市档案馆成立于1960年10月，现有馆藏档案402个全宗，档案156658卷（盒）、资料6014册，馆藏档案内容丰富，门类齐全，载体多样。楚雄市档案馆2015年1月被国家档案局认定为国家一级综合档案馆。

楚雄市档案馆为6层框架结构，设计使用年限50年。建筑采用“回”字形布局，创造了明确的轴线空间，建筑形体采用分台处理来呼应地势高差，保证了建筑形体的完整，南北侧为6层，东侧为4层，建筑在3层平面衔接。新馆功能分区合理，顺应档案馆向大众开放的趋势，引入公共服务功能来打破传统档案馆内部单调的空间布局形式。大楼一、二层围绕内部花园布置开放式查阅大厅及办公室，为公众提供了良好的视觉环境；三、四、五层为档案库房，自成一区，避免干扰；六层为综合业务用房（功能用房）。自三层开始在建筑四角均设置交通核，便于日常的使用和疏散，南侧入口处还设置了无障碍通道，增强了各个功能区的可达性。

新馆建筑造型现代简约、大气方正，外部形式与内部空间设计相呼应。以玻璃幕墙作为建筑的主形象面，增加了建筑本身与外部环境的通透性，体现了档案馆本身的开放性特质。立面的条窗点缀竖向杆件，丰富了建筑的层次和形式，彰显了文化建筑的特色。

Established in October 1960, Chuxiong Archives has a collection of 402 fonds, with 156,658 archival files (cassettes) and 6,014 volumes of documents, all in rich contents, complete range of classifications and various carriers. It was recognized as a National Class-1 General Archives by the National Archives Administration in January 2015.

Chuxiong Archives is a 6-story framework structure building designed with a service life of 50 years. The building's layout in the shape of Chinese character " 回 " forms creates axis spaces. Sub-stages are designed to echo the height difference of the terrain to maintain the integrity of the building's form. It has 6 floors on the north and south parts, and 4 on the east, with planar connection provided on 3F. The new Archives features reasonable functional zoning, and, conforming to the trend that archives are opened to the citizens, public service functions are introduced to break the monotonous, boring spatial layout of traditional archives. On 1F and 2F of the building there are open Reference Hall and offices arranged around the inner garden, offering a good visual environment for the public; 3F, 4F and 5F accommodate archives repositories, which constitute a self-contained zone to avoid interference; 6F is for comprehensive business rooms (functional rooms). Traffic cores are designed at the four corners of 3F and above floors for daily use and evacuation; in addition, a wheelchair accessible passage is provided at the entrance in the south for improved accessibility of each functional area.

In a modern, simple, grand and upright architectural style, the new building has an external form working in concert with its interior space design. Glass curtain walls as the main image facade increase the transparency between the building and the external environment, reflecting the openness of the Archives. The vertical long windows on the facade and interspersed vertical bars enrich the facade's gradations and form, highlighting the characteristics of the Archives as a cultural architecture.

1 档案馆外景 / The Archives Exterior

档案馆

2 | 3
4
5 | 6

2 档案查阅大厅 1 / Archives Reference Hall 1
3 档案查阅大厅 2 / Archives Reference Hall 2
4 入口大厅 / Entrance Lobby
5 库房 / Repository
6 展厅 / Exhibition Hall

澜沧拉祜族自治县档案馆

Archives of Lancang Lahu Autonomous County

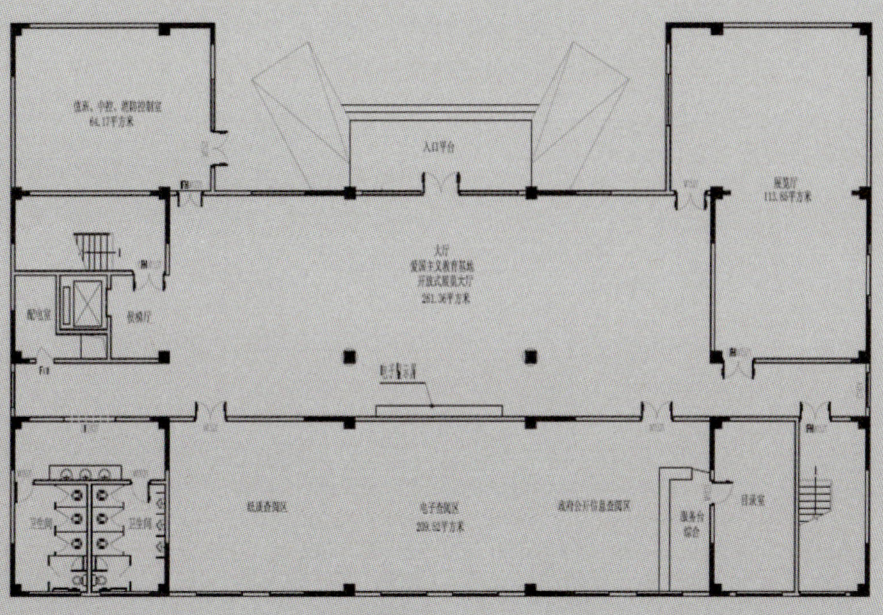

一层平面图 / Ground Floor Plan

占地面积：3333 m²
建筑面积：4643 m²
库房面积：2389 m²
层　　数：6 层
设计单位：昆明理工大学设计研究院
建成时间：2016 年 10 月

Site area: 3,333m²
Floor area: 4,643m²
Repository area: 2,389m²
Number of floors: 6 floors
Time of completion: October 2016

澜沧拉祜族自治县档案馆现有馆藏档案达 97 个全宗，47466 卷，以件为保管单位的档案 298495 件，声像档案 579 盘，照片档案 13570 张，馆藏资料 9398 册。馆藏特色档案中有关拉祜档案反映了新时期的拉祜文化传承人的基本情况，具有拉祜民族风情等特点。

澜沧拉祜族自治县档案馆新馆坐落于行政办公区，环境优雅，紧邻城市干道，交通便捷，周边基础设施完善，具有良好的区位位置和交通优势。新馆平面采用 U 形半围合的布局方式，具有明确的导向性，增强了建筑的可识别性。建筑功能分区紧凑、流线清晰，一层为对外服务用房，主要为展厅和查阅区，其中开放式的爱国主义教育基地展厅与入口门厅相结合，增强了建筑的利用率；二层南侧开间相对较大，主要布置了技术服务用房，包括数字加工服务器机房、收集整理室等，北侧内部开间较小，灵活布置了办公用房；三层以上主要布置库房等。

新馆建筑风格以拉祜族民族特色为主要基调，提取当地文化特色与肌理，抽象为建筑元素运用于立面设计上。立面设计采用主入口内凹两侧对称突出的三段式做法，整体以石材、仿木结构、灰瓦为主，局部点缀民族图腾雕饰、木结构檀条、柱梁穿插等，赋予了新馆澜沧特有的地域风格。建筑以灰白色调为主体建筑的色彩主基调，局部雕饰地方建筑符号。建筑造型清新美观且富有民族特色，矗立于城市建筑中，与周边环境更加和谐统一，传达出档案馆记载和传承着拉祜族悠久历史文化精神的信息。

1 档案馆外景 / The Archives Exterior

Lancang Archives has a collection of 97 fonds, with 47,466 files of archives, as well as 298,495 archival items, with 579 audio/visual disks, 13,570 archival photos and 9,398 volumes of documents. The Lahu nationality related archives in the holding provides the basic information of Lahu culture inheritors in the new era, reflecting the characteristics of Lahu ethnic customs .

Located in the county's administrative office area, the new building of Lancang Archives has an elegant environment; being close to the urban trunk roads, it enjoys convenient transportation, complete surrounding infrastructure, and advantageous location as well as transportation. The U-shaped semi-enclosed layout adopted for the new building presents clear orientation and enhances its recognizability. The building has compact functional zones and clearly defined streamlines. 1F accommodates public service rooms, mainly the Exhibition Hall and reference area; the open Patriotism Education Base Exhibition Hall, in combination with the Entrance Hall, improves the utilization rate of the building. The span in the south of 2F is relatively large in size and mainly used for technical service rooms, including the Digital Processing Server Room, Collection and Sorting Room, etc.; the north part has a relatively small span where offices are arranged in a flexible way. 3F and above floors are mainly for repositories and other rooms.

Lahu ethnic characteristics are used as the main tone in the new building's architectural style, with local cultural features and textures transformed into abstract architectural elements utilized in the facade design. A three-part facade design is adopted, where the main entrance is recessed and both sides protrude symmetrically; on the main materials of stone, synthetic wood structure and gray tiles, ethnic totem carving, wood structure purlins, and interspersed columns and beams, etc. are used as local embellishments, endowing the building with unique local style of Lancang. The mainly building takes a gray and white color tone, with local architectural symbols carved at specific positions. Standing among urban architectures, the new Archives has a fresh, beautiful appearance and rich national characteristics that make it more harmonious and unified with the surrounding environment, conveying the information that it preserves the records and heritage of the long history and culture of the Lahu people.

2 外立面细部特色 / Facade Details

3 | 4
5 | 6
7

3 会议室 / Conference Room
4 一层大厅 / GF Lobby
5 库房 / Repository
6 档案阅览室 / Archives Reading Room
7 展厅 / Exhibition Hall

陇川县档案馆

Longchuan Archives

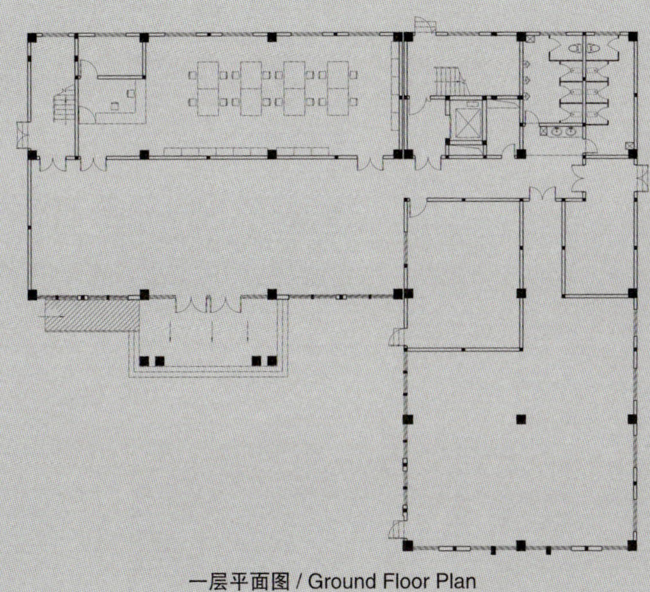

一层平面图 / Ground Floor Plan

占地面积：745m²
建筑面积：3602m²
库房面积：1800m²
层　　数：5层
设计单位：云南省保山市科丹建筑设计有限公司
建成时间：2015年5月

Site area: 745m²
Floor area: 3,602m²
Repository area: 1,800m²
Number of floors: 5 floors
Time of completion: May 2015

陇川县档案馆成立于1961年12月，现拥有馆藏全宗155个，馆藏档案32823卷又42555件，资料5569卷（册）。

陇川县档案馆新馆位于陇川县城章凤镇利民路，紧靠陇川县司法局，毗邻陇川县水利局和陇川县人民医院，地理位置优越，于2012年通过立项，2015年5月全部完工。新馆平面采用L形的布局方式，在转角处布置交通核和辅助用房，两侧布置主要功能用房，功能分区合理，避免了流线的交叉干扰。其中一层主要布置人流量较大的公共服务区，包括展厅、阅览室等；二层北侧布置技术用房和内部办公用房，通风采光较好，东侧大空间布置展厅；三~五层布置库房和资料室等。

新馆建成后，不断完善馆内设施设备，安装档案密集架385m^3，配备了除湿机、空调、防盗安全监控设备、消防栓、干粉灭火器、七氟丙烷自动感应灭火系统，为档案的安全保管提供了安全保障。

档案馆形体呈L形，面向城市道路打开，围合出了入口广场，形成了建筑与环境之间的过渡，与环境融合较好。建筑采用坡屋顶设计，错落变化，营造了丰富的“第五立面”。建筑主入口的台阶、柱廊，给人庄重平稳的感受，入口后的两组立柱延伸至屋顶，强调了建筑的竖向线条。立面的条窗在入口处得到了放大，加大了建筑的窗墙比，保证了建筑内部的采光。建筑立面整体简洁明快、局部设计在遵循整体风格统一的基础上又有变化，符合档案馆建筑自身性质。

1 档案馆外景 / The Archives Exterior

Established in December 1961, Longchuan Archives now has in its collection 155 fonds, with 32,823 files and 42,555 items of archives, as well as 5,569 files (volumes) of documents.

The new building of Longchuan Archives is seated in Limin Road, Zhangfeng Town of Longchuan County, an advantageous location close to the County Justice Bureau, adjacent to the County Water Resources Bureau and County People's Hospital. The construction was approved in 2012 and completed in May 2015. An L-shaped planar layout is adopted, with traffic cores and auxiliary rooms arranged at the corners, and main functional rooms on both sides to create reasonable functional zones and avoid cross-interference of streamlines. 1F of the building mainly accommodates public service area with large flows of people, including the exhibition hall and reading room; on 2F technical room and internal offices are arranged in the north part to enjoy the favorable ventilation and lighting conditions, and the large space in the east part is for exhibition halls; 3F to 5F are used for repositories and documentary archives room.

After the new building was completed, more facilities and equipment have been added, including compact shelving that occupies $385m^2$, dehumidifiers, air conditioners, anti-theft security monitoring devices, fire hydrants, dry powder fire extinguishers, and heptafluoropropane automatic induction fire extinguishing systems, ensuring the safety of archival custody.

The Archives takes an L-shaped appearance and opens to the urban roads, with the enclosed entrance square serving as a transition and ideal integration between the building and its ambient environment. The building adopts a pitched roof design in picturesque disorder which creates a rich "fifth facade". The steps and colonnade at the main entrance of the building presents a feeling of solemnness and smoothness. Two sets of stand columns behind the entrance extend to the roof, emphasizing the vertical lines of the building. The vertical long windows on the facade are enlarged at the entrance, increasing the window-to-wall ratio and guaranteeing the lighting inside. The building's facade's overall style, which is simple and clear, and the local design, which seeks changes in the uniform overall style, both are consistent with the building's nature as an archives.

2 | 3
4 | 5

2 库房 1 / Repository 1
3 档案查阅室 1 / Archives Reference Hall 1
4 档案查阅室 2 / Archives Reference Hall 2
5 库房 2 / Repository 2

勐腊县档案馆

Mengla Archives

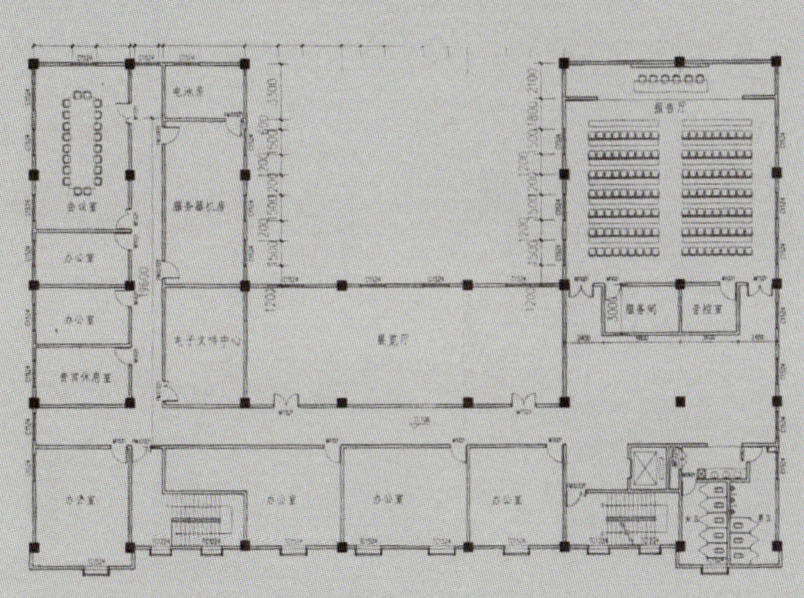

一层平面图 / Ground Floor Plan

占地面积： 3333m²
建筑面积： 4513m²
库房面积： 2149m²
层　　数： 4 层
设计单位： 云南省城乡规划设计研究院
建成时间： 2014 年 10 月

Site area: 3,333m²
Floor area: 4,513m²
Repository area: 2,149m²
Number of floors: 4 floors
Time of completion: October 2014

勐腊县档案馆成立于1981年10月15日，现有馆藏135个全宗，36300卷（册）档案。

勐腊县档案馆新馆坐落勐腊县新城行政中心内，项目为4层框架结构，设计使用年限50年。勐腊县档案馆设计按照公共档案馆功能需求，新馆采用U形平面，半围合的布局形式将各功能空间有序的布置，而其形成三面围合的空间产生了一定的空间围合感，并且通过打开的一面与城市相互联系，使得整个建筑具有了独立于城市的完整空间形态，并且创造出了明确的建筑轴线。建筑一层设置有查阅室、展厅、监控室等，充分发挥档案馆的服务功能，体现档案工作的文化底蕴和内涵；二、三层主要为档案库房、档案消毒室、征收整理室、档案修复室等；四层为办公室、报告厅、小展厅、会议室等，采光通风较好。

在设计理念、风格特色方面，勐腊县档案馆设计既考虑国家现行建筑规范标准、技术合理性，又结合了勐腊地理、气候特点、民族特色。立面兼顾传统与现代，以白色外墙面为主，竖向的开窗保持一定的韵律感，既可以削弱行政办公建筑过于严肃的形象，增加建筑的亲和力，同时又维持了行政办公建筑的庄重稳定身份。建筑整体方正统一，局部处点缀傣族传统文化元素，如顶部具有傣家传统特色的橘色三重顶，腰线处配有傣族特色金腰带条纹及象征图腾的金象浮雕，呈现出新颖、美观特点，是一座具有现代气息又有浓郁民族特色的新型综合档案馆。

Established on October 15, 1981, Mengla Archives now has 135 fonds, with 36,300 files (volumes) of archives in its holding.

Located in the Administrative Center in the new town of Mengla County, the new building of Mengla Archives is a four-stories framework structure designed with a service life of 50 years. The new Archives is designed in accordance with the functional requirements for public archives. It adopts a U-shaped layout where the functional spaces are arranged orderly in the semi-enclosed layout, creating a proper enclosed space from the three enclosing sides, and linking with the urban spaces through the open side; this gives the whole building a complete spatial form independent of the city, and creates a clear architectural axis. On 1F the Reference Room, Exhibition Hall, Monitoring Room, etc.,

1 档案馆外景 / The Archives Exterior

are arranged, giving full play to the service functions of the Archives and reflecting the cultural heritage and connotations of archival work; 2F and 3F mainly accommodate Archives Repositories, Archives Disinfection Room, Collection and Sorting Room, Archives Repair Room, etc.; and on 4F with good lighting and ventilation conditions, there are offices, Lecture Hall, Small Exhibition Hall, Conference Room, etc.

In terms of design concepts, styles and characteristics, the design of Mengla Archives considered the current national building codes, standards and technical rationality in combination with the local geographical, climatic and ethnic characteristics of Mengla. The facade, giving attention to both and traditional modern , mainly consists of white exterior walls. The vertical windows on the facade present a certain sense of rhythms, which to a certain extent make the building's image as an administrative office architecture less rigid, and also maintain its solemnness and stableness. The building's overall structure is unified, upright and foursquare, with traditional cultural elements of the Dai ethnic group used as embellishment at some places. For example, the orange triple top patterns as a unique Dai ethnic group feature on the building's top, as well as the typical gold-colored belt stripes and golden elephant reliefs symbolizing the totem in Dai culture used at the band course; all these render is a novel, beautiful new comprehensive archives with both modern style and strong national characteristics.

2
3
4 | 5

2 档案阅览室 / Archives Reading Room
3 会议室 / Conference Room
4 库房 / Repository
5 门厅 / Lobby

巍山彝族回族自治县档案馆

Archives of Weishan Yi and Hui Autonomous County

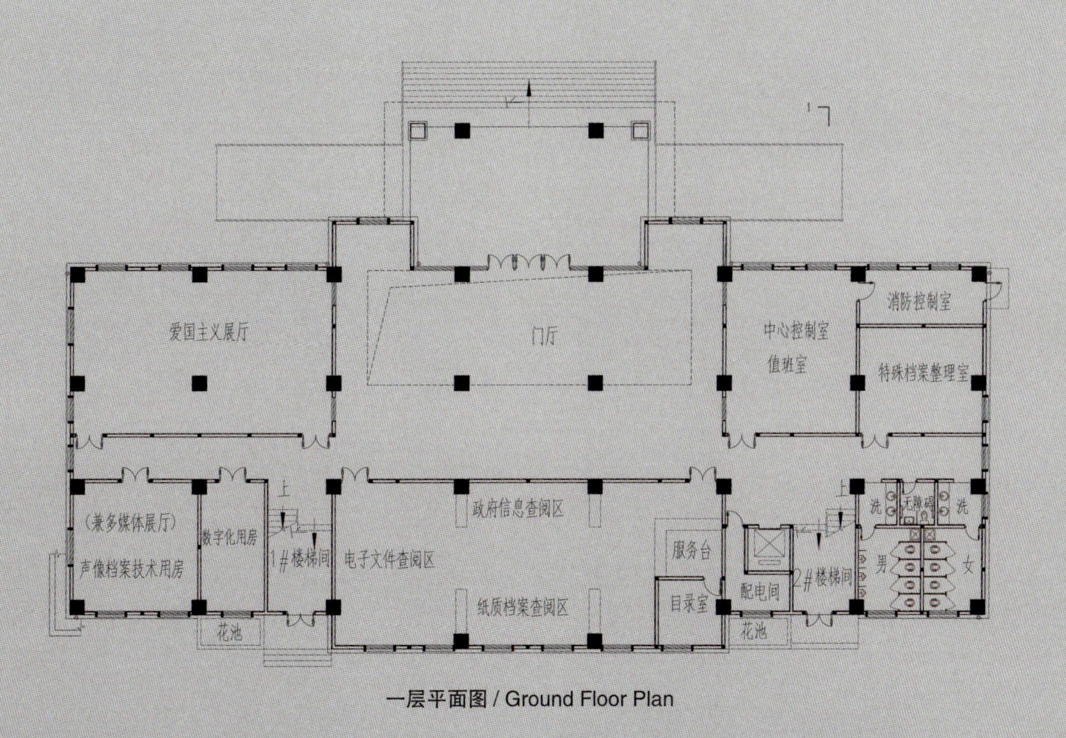

一层平面图 / Ground Floor Plan

占地面积： 3086m²
建筑面积： 4769m²
库房面积： 2381m²
层　　数： 5层
设计单位： 云南博超建筑设计有限公司
建成时间： 2014年2月

Site area: 3,086m²
Floor area: 4,769m²
Repository area: 2381m²
Number of floors: 5 floors
Time of completion: February 2014

巍山彝族回族自治县档案馆成立于1959年6月，现有馆藏档案106个全宗，档案40288卷，其中年代最早的档案为明末清初的《续修蒙化直隶厅志》。

巍山彝族回族自治县档案馆新馆位于南诏镇文献新区，于2013年3月开工，2014年2月竣工。平面呈“一”字形排布，节约建造成本，与地形结合较好。档案馆将面积需求较大的开放性展览阅览空间置于一、二层，便于公众使用，包括开放档案阅览室、电子档案阅览室、政府公开信息阅览室、缩微档案阅览室、音像档案阅览室、爱国主义教育展厅等；将同样面积要求较大的库房置于三、四层，便于其使用的独立性，包括重大活动档案陈列室和保管各种载体档案的库房；顶层私密性较好，布置多功能报告厅和内部办公用房，内部安防设施完备。平面布局紧凑、流线清晰，方便公众使用。

新馆为白墙灰瓦的新中式风格，整体建筑分为墙基、墙身、屋顶三个基本部分，并置的关系强化了建筑的传统渊源。其中屋顶采用中国传统的四坡屋顶，但并不强调瓦的存在，因此出挑尺度不大。立面采用三段式，在入口处打断下部檐口，丰富了立面层次，开窗结合功能需求在同一个模数中寻求变化。墙身的垂直线条元素体现了对传统木构的模仿。两个交通核对称突出屋顶，进一步加强了建筑的庄重和质朴的特质，既与传统古城建筑相协调，又与自然相和谐，与周围建筑相呼应，反映了以人为本和可持续发展的理念。

1 档案馆外景 / The Archives Exterior

Established in June 1959, Weishan Archives has 106 fonds and 40,288 files of archives in its holding, the earliest among which is Renewed Records of Menghua Zhili Hall.

The new building of the Archives is located in the Wenxian New District of Nanzhao Town. The construction was started in March 2013 and completed in February 2014. It adopts an I-shaped planar layout, saving construction costs and adapting to the terrain conditions. The open exhibition and reading spaces which require large areas are arranged on the first and second floors for convenient public use, including the Open Archives Reading Room, Electronic Archives Reading Room, Governmental Information Reading Room, Micro Archives Reading Room, Audio-visual Archives Reading Room, Patriotism Education Exhibition Hall, etc.; repositories including the Major Events Archives Exhibition Room and repositories for preserving archives of various carriers, which also need large areas, are arranged on 3F and 4F for separation and independent spaces; the top floor, which is more private and therefore the Multi-functional Lecture Hall and internal offices are arranged there with security facilities equipped inside. The planar layout is compact with clear streamlines for convenient use by the public.

Taking a new Chinese style with white walls and grey tiles, the entire new building is divided into three major parts, namely the wall foundation, wall body and roof, the juxtaposition of which highlighting the traditional origins of the building. The traditional Chinese four-slope roof is adopted, without emphasizing the existence of tiles and therefore making only small overhang. On the three-parts facade, the lower eaves is interrupted at the entrance to enrich the facade gradations; window openings showcase the pursuit of variation considering functions in the same modulus. The vertical lines on the walls reflect the imitation of traditional wood structure. The two traffic cores symmetrically protrude from the roof, further strengthening the solemnness and plainness of the building. It is in harmony with the traditional ancient town architectures and united with the nature, echoing the surrounding buildings and reflecting the people-oriented, sustainable development concepts.

2
3 | 4
5 | 6

2 档案阅览室 / Archives Reading Room
3 展厅 1 / Exhibition Hall 1
4 展厅 2 / Exhibition Hall 2
5 展厅 3 / Exhibition Hall 3
6 一层大厅 / GF Lobby

永德县档案馆

Yongde Archives

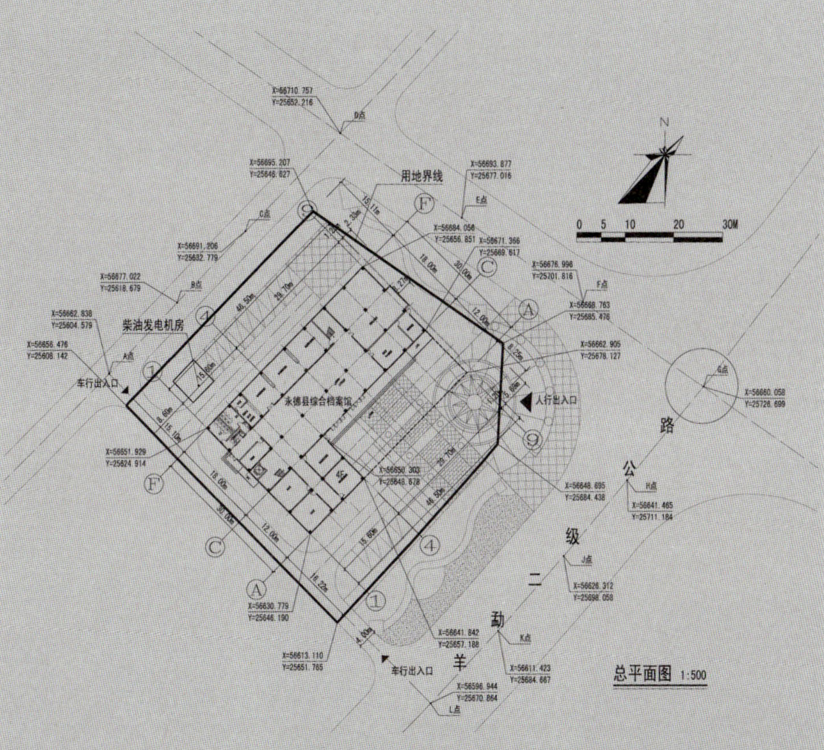

总平面图 / Master Plan

占地面积：3833m²
建筑面积：5056m²
库房面积：2587m²
层　　数：5 层
设计单位：云南永德建筑工程有限公司
建成时间：2015 年 10 月

Site area: 3,833m²
Floor area: 5,056m²
Repository area: 2,587m²
Number of floors: 5 floors
Time of completion: October 2015

永德县档案馆于 1959 年 11 月成立，现有馆藏全宗 142 个，档案资料总数量 83547 卷（册、件）。

永德县档案馆新馆位于城南开发区，设计使用年限为 50 年。地上建筑 5 层，建筑采用 L 形布局，功能分区明确，流线合理，其中档案库房面积 $2587m^2$，可满足未来 50 年档案保管需求；对外服务用房面积 $1380m^2$，能够满足档案对外利用服务、政府公开信息查阅的法定场所需求，满足国家档案馆作为安全保管党和国家重要档案的基地、爱国主义教育基地需求；档案业务和技术用房面积 $357m^2$，能够满足档案资料保管和利用档案的功能要求；办公室用房面积 $210m^2$，能够满足日常办公需求；附属用房面积 $521m^2$，能够满足综合档案馆供电、供水、消防、节能、环保的需求。

建筑形体半围合成的绿化广场将城市空间延绵入场地之内，正立面两层高的门廊，通透的入口门厅都与环境有着很好的互动效应，吸引和方便公众，更好地发挥档案馆的开放服务功能。建筑立面采用三段式布局，基座为灰色石材，厚重大气；屋身为浅黄色，立面开洞的竖向处理形成强烈的韵律感，在一定程度上淡化了行政办公建筑的严肃形象；屋顶采用中国传统的坡屋顶，出挑深远，塑造了新中式的建筑风格。新馆建筑不仅体现了中华传统特色与现代建筑的有机结合，也丰富了城市界面，提高了城市地位。

Established in December 1959, Yongde Archives now has in its collection 142 fonds, with a total of 83,547 files (volumes, items) of archives.

The new building of Yongde Archives is located in Chengnan Development Zone, and is designed with a service life of 50 years. It is a five-story architecture aboveground with an L-shaped layout and clearly defined functional areas as well as reasonable arranged streamlines. The Archives has repositories occupying $2,587m^2$, which can meet the needs of archival storage in the next 50 years; the public service space is $1,380m^2$, up to the requirements for serving as an archival information service center, a disclosed prevailing document access and use center, and a legal venue for governmental information reference, as well as the

1 档案馆外景 / The Archives Exterior

requirements for a national archives to act as the base for safely preserving Party and national important archives, and a patriotism education base; the area for archives business and technical rooms is $357m^2$, which can satisfy the functional demands for archives custody and use; the office space of $210m^2$ is adequate for daily needs; and the area of auxiliary rooms, $521m^2$ is enough to reach the standard for a general archives in terms of power supply, water supply, fire protection, as well as energy saving and environmental protection.

A green square is shaped by the semi-enclosure of the building's structure, extending the urban spaces into the site. The two-story porch on the front elevation and the pass-through entrance hallway not only echo the ambient environment perfectly but also attract and facilitate the public, making it easier to exert the Archives' public service functions. With a three-part layout, the building facade has a thick and heavy foundation made of gray stone; the building's main body is light yellow, with vertical openings on the facade forming a strong sense of rhythms, which to a certain extent makes the building's image as an administrative office architecture less rigid; the roof, in a traditional Chinese pitched style and far-reaching, creates a new Chinese architectural style. The new building not only organically combines traditional Chinese characteristics with modern architectural style, but also enriches the urban interface and enhances the city's status.

2
3 | 4
5 |

2 入口大厅 / Entrance Lobby
3 档案查阅室 / Archives Reference Room
4 外立面局部 / Part of Elevation
5 库房 / Repository

墨脱县档案馆

Medog Archives

Tibet Autonomous Region

主立面图 / Main Elevation Plan

新馆 / New

占地面积：3000m^2
建筑面积：895m^2
库房面积：545m^2
层　　数：6 层
设计单位：西藏金潮工程设计有限公司
建成时间：2014 年 5 月

Site area: 3,000m^2
Floor area: 895m^2
Repository area: 545m^2
Number of floors: 6 floors
Time of completion: May 2014

墨脱县档案馆现有馆藏文书档案1.2万卷（件、册），另保存有其他专业档案130卷，科技档案80盒，照片档案857张。

墨脱县档案馆新馆坐落于西藏林芝墨脱县人民政府大院内，与财政局、国土局、民政局、工商局、住建局等单位位于同一栋大楼中。档案馆部分一楼为值班室、办公室、档案阅览室和复印室等，二、三楼均为档案库房。

墨脱县档案馆外立面设计为典型的藏式风格，白色墙面与红色屋檐相结合，在西藏地区，白色与红色分别代表天上与地上。整体设计理念立足本土门巴、珞巴族文化，在充分发挥档案收集、整理保护的基础上加强了民族文化的收集、保护、研究、交流，展示门珞文化价值。档案馆设计风格将门珞传统文化与现代建筑特征融为一体，充分体现了各民族团结与和谐统一。

Medog Archives has in its collection of 12,000 files (items, volumes) of administrative archives, as well as 130 other specialized archives, 80 cassettes of scientific and technological archives and 857 archival photos.

The new building of Medog Archives is in the People's Government compound of Medog County of Tibet, in the same building as the county's Finance Bureau, Land and Resources Bureau, Civil Affairs Bureau, Industry and Commerce Administration, and Housing and Construction Bureau. In the archives part of the building, 1F accommodates the Duty Room, Offices, Archives Reading Room and Copy Room, and repositories are on 2F and 3F.

The exterior facade of the Archives is designed in a typical Tibetan style, with white walls used in combination with red eaves, which colors in Tibet represent the heavens and the earth respectively. The overall design philosophy is based on the local cultures of the Moinba and Lhoba nationalities, which, on the basis of giving full play to the archives' functions including archival collection, sorting and preservation, strengthens the collection, protection, research and exchange of national cultures, and showcases the values of the Moinba and Lhoba cultures. Traditional Moinba and Lhoba cultures and modern architecture characteristics are integrated into the archives' design style, fully reflecting unity and harmony among all ethnic groups.

1 档案馆外景 / The Archives Exterior

2
3 | 4
5 | 6

2 库房 / Repository
3 档案整理室 / Archives Filing Room
4 库房外部 / Repository Exterior
5 入口 / Entrance
6 会议室 / Conference Room

合阳县档案馆

Heyang Archives

Shaanxi Province

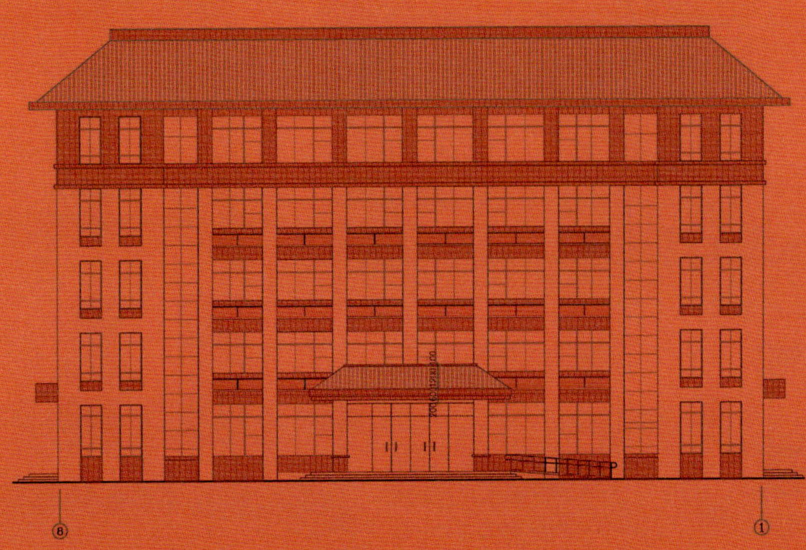

正立面图 / Front Elevation Plan

旧馆 Old

New 新馆

占地面积： 2520m²
建筑面积： 3326m²
库房面积： 1400m²
层　　数： 主楼 5 层，配楼 1 层
设计单位： 陕西建协设计研究院
建成时间： 2018 年 5 月

Site area: 2,520m²
Floor area: 3,326m²
Repository area: 1,400m²
Number of floors: 5 floors for the main building and 1 floor for the wing
Time of completion: May 2018

合阳县档案馆成立于 1958 年 10 月，现有馆藏档案 7 个门类、120 个全宗，126847 卷（件、册），涉及文书档案、会计档案、专业档案、科技档案以及个人档案等各个门类。

合阳县档案馆新馆位于陕西省渭南市合阳县城关街道办事处水车头村西头（凤凰南路金水派出所对面），坐南朝北。新馆设计时充分考虑了档案馆要同时满足保存、保管、研究、展览与教育和文化传播等功能。在平面布局上，既注意到功能的整合利用，又避免了流线的交叉，在最大程度上方便所有人的使用。

档案馆整体设计上立足合阳地域文化，紧扣陕西历史文化名片，采用传统坡屋顶、三段式立面，庄重而不豪华，雅致而不张扬。同时，立面穿插横向的线条，使得建筑造型不仅充满古典韵味，还融入了现代审美的简洁挺拔。整个建筑既有端庄肃穆的氛围，又有新颖别致的巧思。

合阳县档案馆的建成，为广大干部群众及时了解党和政府的方针政策提供了方便，对建设成经济强县和旅游名县发挥了积极的作用。

Established in October 1958, Heyang Archives has in its holding 7 classifications of archives, including 120 fonds and 126,847 files (volumes, items) of archives including administrative archives, accounting archives, specialized archives, scientific and technology archives, and personal archives.

The new building of Heyang Archives, sitting northward, is located in the west end of Shuichetou Village, Chengguan sub-district office, Heyang County, Weinan City, Shaanxi Province (across the road to Jinshui Police Station on the South Fenghuang Road). The new Archives' design takes into full consideration that it shall meet all necessary functions including preservation, custody, research, exhibition and education, cultural communication, etc. The planar layout pays attention to both the integrated utilization of functions and avoidance of streamline intersection, so as to facilitate access and use of everyone to the greatest extent.

The overall design of the Archives is based on the regional culture of Heyang and closely linked to the purpose to make the building a historical and cultural business card of Shaanxi; traditional sloping roof and three-part facade is adopted to create a feeling of solemnness, low profile, elegance and simplicity. At the same time, horizontal lines are inserted into the facade so that the building is both full of classical charms and takes on the simplicity and straightness of modern aesthetics. The entire building is imbued with a dignified, solemn atmosphere as well as novel, ingenious ideas.

The new building of Heyang Archives has provided convenience for the broad masses and cadres to get timely informed of the Party and the governments' principles and policies, and played a positive role in building Heyang into a powerful county in economy and a famous tourist county.

1 档案馆外景 / The Archives Exterior

增强档案法制意识，依法保护利用档案信息。
金 水

2
3
4 | 5

2 一层大厅 / GF Lobby
3 展厅 1 / Exhibition Hall 1
4 展厅 2 / Exhibition Hall 2
5 档案查阅室 / Archives Reference Room

山阳县档案馆

Shanyang Archives

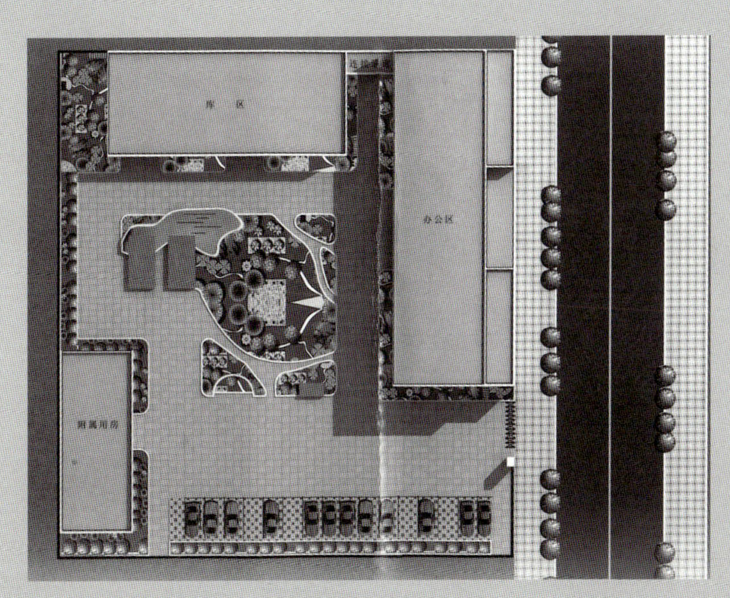

总平面图 / Master Plan

占地面积：3333m^2
建筑面积：4359m^2
库房面积：1482m^2
层　　数：主楼 5 层，配楼 4 层
设计单位：商洛市建筑勘察设计院
建成时间：2014 年 6 月

Site area: 3,333m^2
Floor area: 4,359m^2
Repository area: 1,482m^2
Number of floors: 5 floors for the main building and 4 floor for the wing
Time of completion: June 2014

山阳县档案馆成立于 1958 年 9 月，现有馆藏档案 144 个全宗，共计 64382 卷又 205832 件，文书档案占馆藏档案的 65%，资料 9350 册。2015 年 11 月，山阳县档案馆被评为国家一级档案馆。

山阳县档案馆新馆坐落于山阳县城西河新区。整个建筑平面呈 L 形布置，半围合成的内庭院提供了良好的景观环境及丰富的室外活动空间。外观设计采用竖向柱均匀排布，外墙整体以米白色墙砖饰面，增强了建筑立体感；建筑屋面出檐设计造型为斗栱形，吸收了山阳独特的古代建筑风格，运用现代手法，充分突出陕南文化的连续性和地域特色，彰显国家综合档案馆的“藏史”和公共文化场所的开放服务特点，档案文化标志鲜明；轴线对称的立面设计，使建筑挺拔，渲染档案馆神圣的建筑气氛。

L 形的建筑布局将平面功能分为库房和办公两部分。主楼为办公区，包含技术办公、行政办公及对外接待部分，配楼为档案库房区，两者之间通过走廊连接，既相对独立，又相互联系，保障了档案馆高效、便捷的运作。

1 档案馆外景 / The Archives Exterior

Established in September 1958, Shanyang Archives has a collection of 144 fonds, with 5 classifications including 64,382 files and 205,832 items, among which administrative archives account for 65% of the total holding, as well as 9,350 volumes of documentary archives. In November 2015, Shanyang Archives was awarded National Class-1 Archives.

The new building of Shanyang Archives is located in the Xihe New District of Shanyang County. It takes an L-shaped layout, with a semi-enclosed inner courtyard to provide perfect landscape environment and ample space for outdoor activities. The exterior design includes evenly arranged vertical columns, and exterior walls decorated with beige white tiles to enhance the three-dimensional feeling of the building. The roof has an arched eaves on the roof, a unique ancient architectural style of Shanyang; and modern techniques are also used to fully highlight the continuity and regional characteristics of southern Shaanxi culture, emphasizing the "collection history" of the national archives and the open service features of a public cultural venue, with remarkable archival cultural characteristics; the axis-symmetric facade design makes the building tall and straight, rendering the sacred atmosphere of the architecture.

The L-shaped architectural layout divides the planar functions into two parts, namely repositories and offices. The main building is used as the office area, which includes Technical Offices, Administrative Offices and Reception. The wing building is for repositories, linked with the main building through corridor, so as to keep them relatively independent and interconnected, thus ensuring the efficient, convenient operation of the Archives.

2
3 | 4
5

2 档案查阅室 / Archives Reference Room
3 库房 / Repository
4 展厅 / Exhibition Hall
5 报告厅 / Lecture Hall

长武县档案馆

Changwu Archives

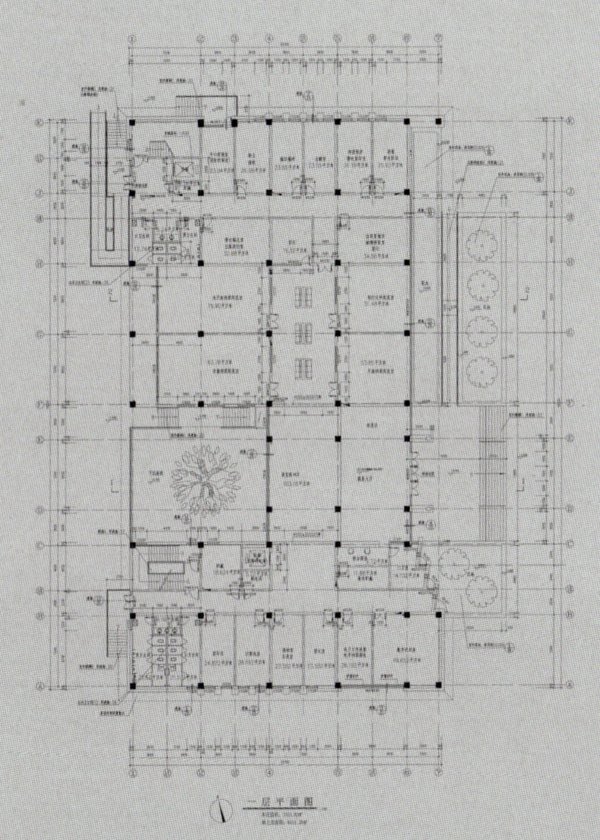

一层平面图 / Ground Floor Plan

占地面积：8026m²
建筑面积：7068m²
库房面积：1993m²
层　　数：地上 4 层，地下 1 层
设计单位：南京城镇规划设计院
建成时间：2016 年 3 月

Site area: 8,026m²
Floor area: 7,068m²
Repository area: 1,993m²
Number of floors: ↑ 4 floors and ↓ 1 level
Time of completion: March 2016

长武县档案馆现存档案资料5个门类179个全宗，共计案卷101648卷（件、册），馆藏构成主要有明清档案、民国时期档案和新中国成立后档案。

长武县档案馆新馆位于长武县新区昭仁大街以南、育才路以北，明珠宾馆以东。建筑总平面采用“凹”字形布局，与长武县图书馆和长武县文化馆形成统一的建筑风格，以浅黄色石材和玻璃幕墙相结合，大气端庄，简洁统一。档案馆正立面虚实相间，以浅黄色石材为实、玻璃幕墙为虚，形成“实中有虚，虚中有实”的立面构成，均衡而又统一。

新馆平面布局合理，一层为展厅、服务大厅及部分办公用房，二层、三层为档案馆库房和技术办公用房，四层为城乡规划馆及休闲活动用房。建筑体块内凹，形成的下沉庭院，营造出良好的景观，使得门厅访客对庭院空间产生兴趣，激活庭院空间，同时沿庭院一侧设置走道空间，增加馆内人员与下沉庭院空间的交流，赋予建筑空间趣味性。

1 档案馆外景 / The Archives Exterior

Changwu Archives has 179 fonds of 5 classifications, totaling 101,648 archival files (items, volumes). The collection mainly consists of archives from the Ming and Qing Dynasties, the period of the Republic of China and the period after the People's Republic of China was founded.

The new archives building is located in the south of Zhaoren Street in the New District of Changwu County, north of Yucai Road and east of Mingzhu Hotel. The concave overall planar layout is in a unified architectural style with the Library and Cultural Center of Changwu County, with light yellow stone combined with glass curtain walls, creating a dignified, grandeur yet concise and uniform style. The front facade combines virtuality with reality, with pale yellow stone representing reality and glass curtain walls indicating virtuality, forming a balanced, unified facade constituted by "virtuality within reality and vice versa".

The building has a reasonable layout with Exhibition Hall, Service Hall and some offices on 1F, Archives Repositories and Technical Offices on 2F and 3F, and Urban and Rural Planning Hall and Recreational Room on 4F. The submerged courtyard formed by the concave masses of the building forms perfect landscape, which attracts visitors at the entrance hall to the courtyard space, and makes the courtyard space lively; a walkway space is arranged along one side of the courtyard to increase the communication between people in the building and the submerged courtyard space and make the architectural space more interesting.

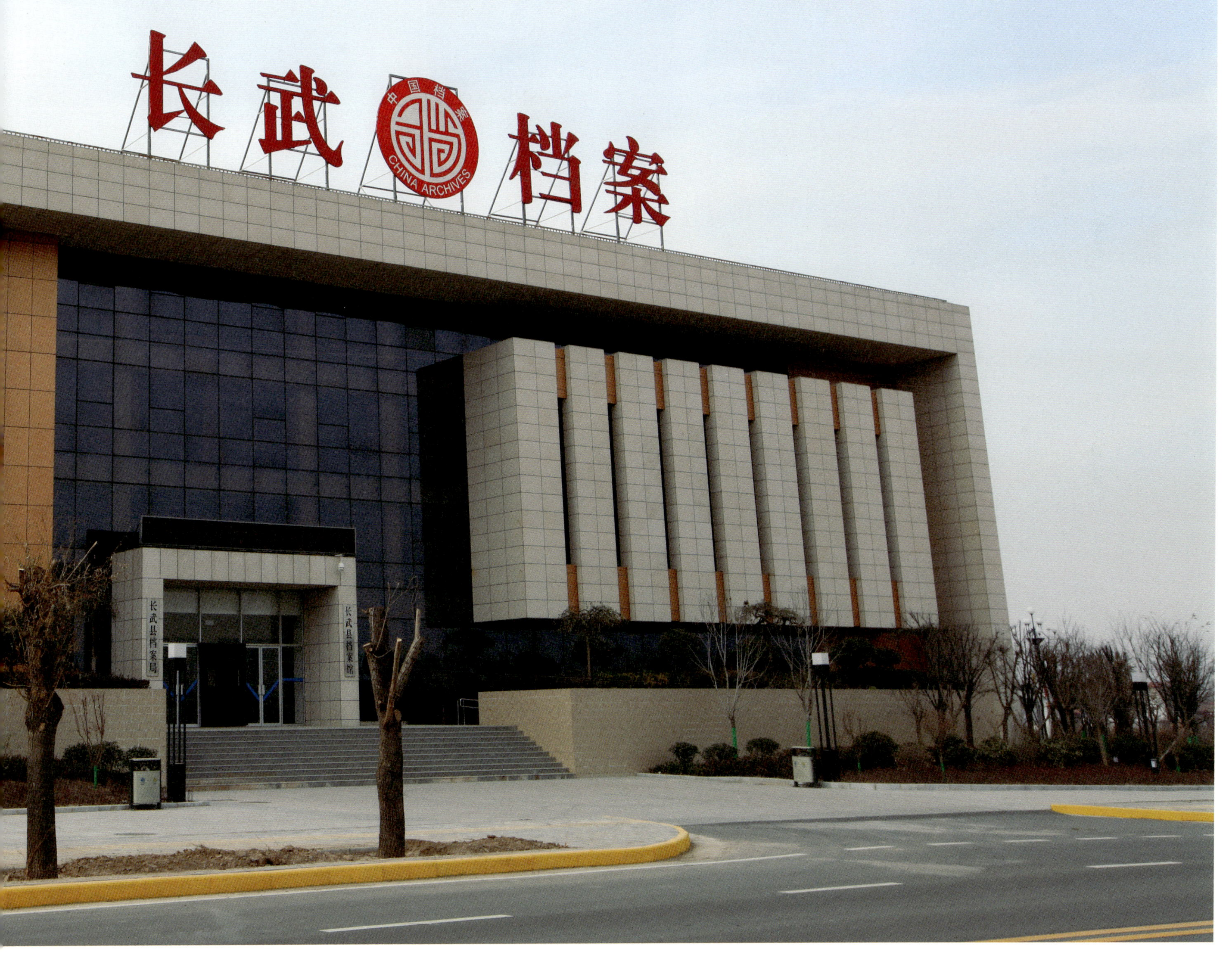

2

3

4 | 5

2 资料室 / Documentary Archives Room
3 会议室 / Conference Room
4 入口 / Entrance
5 库房 / Repository

金昌市金川区档案馆

Jinchuan District Archives, Jinchang

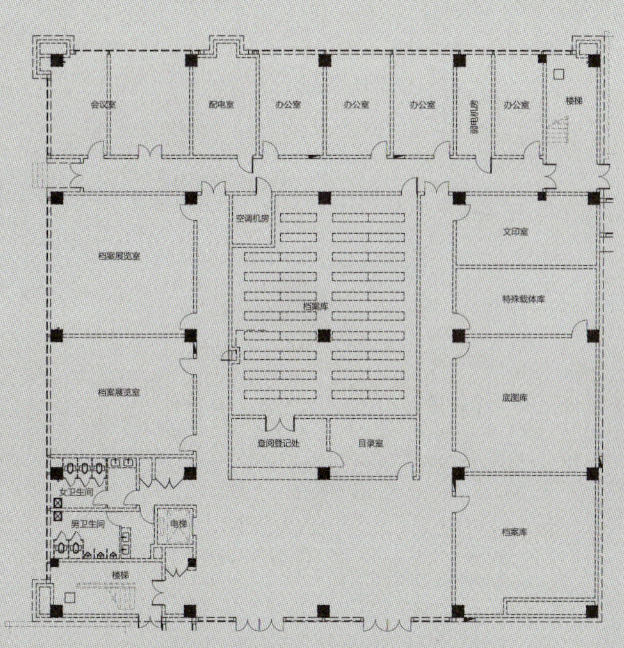

一层平面图 / Ground Floor Plan

占地面积：14500m²
建筑面积：3740m²
库房面积：2100m²
层　　数：3层
设计单位：北京清城华筑建筑设计研究院有限公司
建成时间：2016年8月

Site area: 14,500m²
Floor area: 3,740m²
Repository area: 2,100m²
Number of floors: 3 floors
Time of completion: August 2016

金昌市金川区档案馆成立于1995年4月，现有金川区建区30多年来馆藏的文书档案15593卷128289件，婚姻2289卷22806件，照片档案2300张，荣誉档案302件，实物档案174件。

金昌市金川区档案馆新馆位于金昌市金川区紫金苑景区，与金川区图书馆、文化馆合建，三馆集中于一栋建筑中，既相互独立又紧密联系，彼此间设有共享大厅，为市民呈现出良好的公共性。档案馆位于建筑西北侧，一层、二层中间大空间为档案库房，配备了档案密集架、防磁柜；其他房间围绕库房布置，一层主要为档案展览室和办公室，二层为档案阅览与档案业务、技术及数字处理用房，馆内各种功能设施配备齐全，能够满足今后60年馆藏需要，为金川区档案事业的发展奠定了坚实的基础。

新馆建筑由4个几何体块组合而成，体块之间通过一层进行联系，设计风格简约现代。立面采用大面积的玻璃与石材组合，二层局部镂空处理，富有文化建筑应有的开放气息。

1 档案馆外景 / The Archives Exterior

Established in April 1995, Jinchuan District Archives has a collection of archives through the more than 30 years since the foundation of the Jinchuan District, totaling 15,593 files and 128,289 items of the administrative archives, 2,289 files and 22,806 items of marriage archives, 2,300 archival photos, 302 items of honorary archives, and 174 material objects.

The new building of Jinchuan District Archives is located in the Zijinyuan Scenic Area of Jinchuan District, Jinchang City. It is co-constructed in the same building with the Library and Cultural Center of the District, independent yet closely related to the other two, with a shared hall designed for using in common, showing their public image to the citizens. The Archives is located on the northwest side of the building, with the large spaces in the middle of 1F and 2F used for archives repositories equipped with compact shelving and anti-magnetic cabinets. Other rooms are arranged around the repositories. 1F mainly accommodates Archives Exhibition Rooms and offices, and 2F is for archives reading and archives business, technical and digital processing rooms. The building is fully equipped with various functional facilities to meet the holding requirements for the next 60 years, laying a solid foundation for the development of the archives business in Jinchuan District.

The new building is composed of four geometric masses, which are connected at 1F and designed with a simple yet modern style. The facade is made up of a large area of glasses and stone, and 2F adopts partially hollow-out to create the openness of cultural architectures.

2
3 | 4
5

2 档案查阅大厅 / Archives Reference Hall
3 档案整理室 / Archives Filing Room
4 库房 / Repository
5 数字化室 / Digital Processing Room

刚察县档案馆

Gangcha Archives

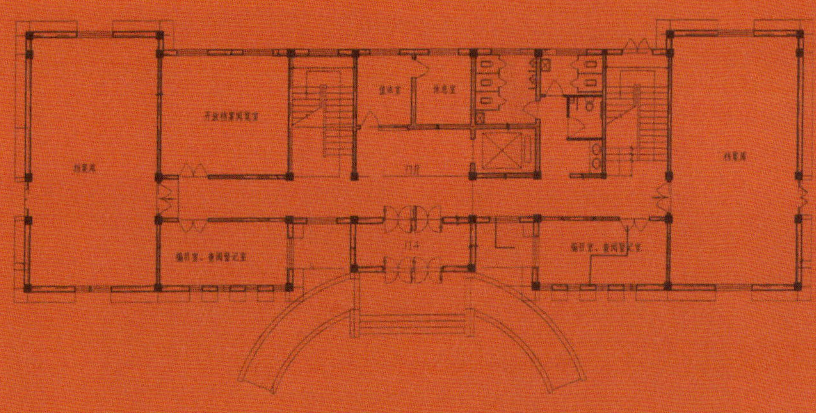

一层平面图 / Ground Floor Plan

占地面积：1500m²
建筑面积：1398m²
库房面积：780m²
层　　数：主楼 3 层，配楼 1 层
设计单位：青海煤炭设计院
建成时间：2012 年 10 月

Site area: 1,500m²
Floor area: 1,398m²
Repository area: 780m²
Number of floors: 3 floors for the main building and 1 floor for the wing
Time of completion: October 2012

刚察县档案馆成立于1958年3月，馆藏档案丰富、门类齐全，共保存58个全宗，各类纸质档案资料共计7.47万卷（册），2013年10月晋升为国家二级档案馆。

刚察县档案馆新馆位于县新城区行政中心区域内，独立建设，与县政府综合办公大楼相邻，周边环境静谧，交通便利。大楼高12m，由主体三层和屋顶装饰建筑组成，馆内功能布局合理，设有档案库房、展厅、档案查阅室、整理室、接收室、目录室、资料室、机房等。档案查询、阅览、登记室与档案库位于一层，方便档案运送与公众使用；二层是档案技术用房与数字化用房；展厅、办公、会议用房位于三层，拥有开阔的视野。

新馆建筑厚重内收的墙体，规律的开窗，红、黄、白色调的墙面均体现了浓郁的藏式文化风格，在高原日光的照射下格外耀眼，是刚察县标志性建筑之一。

1 档案馆外景 / The Archives Exterior

Set up in March 1958, Gangcha Archives has a rich collection covering all classifications, with 58 fonds, and a total of 74,700 files (volumes) of paper documentary archives. In October 2013, it was upgraded to a national class-2 archives.

The new building of Gangcha Archives is located in the Administrative Center of the new urban area, built independently near the county government's comprehensive office building, with a quiet, convenient environment surrounding it. Designed in 12m high and consisting of 3 floors of the main building as well as roof decoration, the building has a reasonable functional layout with rooms including the Archive Repositories, Exhibition Hall, Archives Reference Room, Sorting Room, Accession Room, Catalogue Room, Documentary Archives Room and Computer Room. Archival Referencing, Reading, Registration Rooms and Archives Repositories are located on 1F to facilitate archival transportation and use by the public; 2F accommodates Archives Technical Rooms and Digital Rooms; the Exhibition Hall, Offices and Conference Rooms are located on 3F with an open view.

The massive, inward walls of the new building, with regularly arranged windows and walls in a color tone of red, yellow and white, reflect the typical Tibetan culture style, standing out with special brightness under the sunlight on the plateau. It is one of the landmark architectures of Gangcha County.

2
3
4 | 5

2 展厅 / Exhibition Hall
3 档案查阅室 / Archives Reference Room
4 库房 / Repository
5 目录室 / Catalogue Room

盐池县档案馆

Yanchi Archives

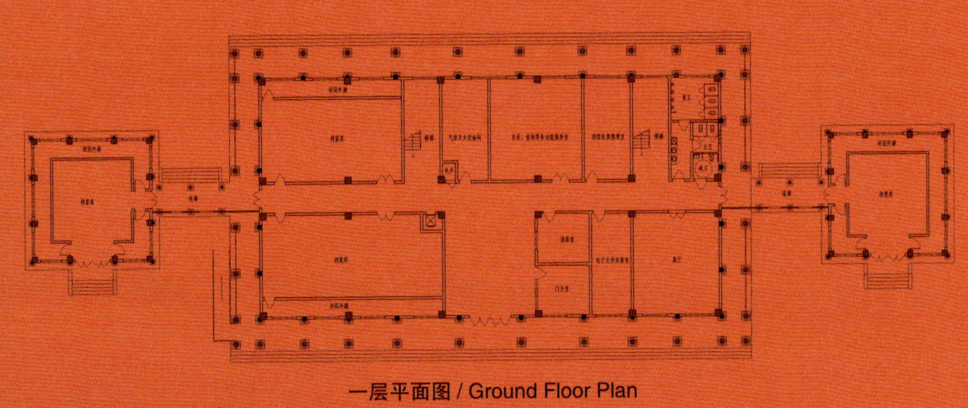

一层平面图 / Ground Floor Plan

占地面积： 5286m²
建筑面积： 2224m²
库房面积： 916m²
层　　数： 主楼 2 层，配楼 1 层
设计单位： 西安长安大学工程设计研究院有限公司
建成时间： 2018 年 12 月

Site area: 5,286m²
Floor area: 2,224m²
Repository area: 916m²
Number of floors: 2 floors for the main building and 1 floor for the wing
Time of completion: December 2018

盐池县档案馆成立于1959年2月，现有各类档案全宗及全宗群201个，馆藏档案折算10万余卷（件），排架长度885m，馆藏资料9000余册。2014年成功创建国家二级档案馆。

盐池县档案馆新馆位于县城东北角，处于仿古建筑文化旅游区域。建筑最大亮点是采用了中国仿古建筑外观，充分结合了中国传统文化气息，做到了档案馆建筑“动”与“静”的结合，醒目的中国红挑檐和屋顶与青砖有机结合，更赋予了档案馆文化建筑特有的底蕴与灵气。

建筑地上两层，功能布局集约紧凑，主要由档案保存、档案利用、档案加工管理与办公四个部分组成。具体包括档案库房、档案展厅、阅览室、业务技术用房、档案数字化与会议办公用房。其中库房内全部安装了智能、手动密集架、恒温恒湿及视频、红外、监控报警设备，库区均设置了封闭走廊，保证库房恒温恒湿。

盐池县档案馆新馆已成为盐池县标志性建筑之一，为讲述县城悠久的历史和丰富的人文添上了重重的一笔丹青。

Established in February 1959, Yanchi Archives has 201 fonds and fond complexes of different classifications, with a total collection that equals more than 100,000 files (items) of archives; it also has shelvings 885m long and over 9,000 volumes of documentary archives. In 2014, it was successfully upgraded to a class-2 national archives.

The new building of Yanchi Archives is located in the northeast corner of the county, rightly in the antique architectural and cultural tourism area. The most attracting part of the building is its appearance of Chinese antique architectures, which fully employs the traditional Chinese cultural elements, and combines "dynamics" and "stillness" of archives buildings. The striking Chinese red overhanging eaves and roofs, in harmony with the grey bricks, endow the Archives with unique connotations and charms of archives as cultural architectures.

The two above-ground floors of the building are arranged in a compact layout, mainly consisting of four parts, i.e. archival storage, archival access and use, archival processing management and offices. Specifically, there are Archives Repositories, Archives Exhibition Halls, Reading Rooms, Business Technical Rooms, Archives Digitization and Conference/Office Rooms. All the repositories are furnished with intelligent and manual compact shelving, constant temperature and constant humidity and video, infrared monitoring and alarm equipments; in the repository areas enclosed corridors are arranged to ensure constant temperature and constant humidity.

The building has become one of the landmark buildings of Yanchi County, adding a significant brushstroke to the county's age-old history and rich humanistic culture.

1 档案馆外景 / The Archives Exterior

2

3

4

2 档案查阅大厅 / Archives Reference Hall

3 库房 / Repository

4 档案馆外景 / The Archives Exterior

库尔勒市档案馆

Korla Archives

新疆

Xinjiang Uygur Autonomous Region

维吾尔自治区

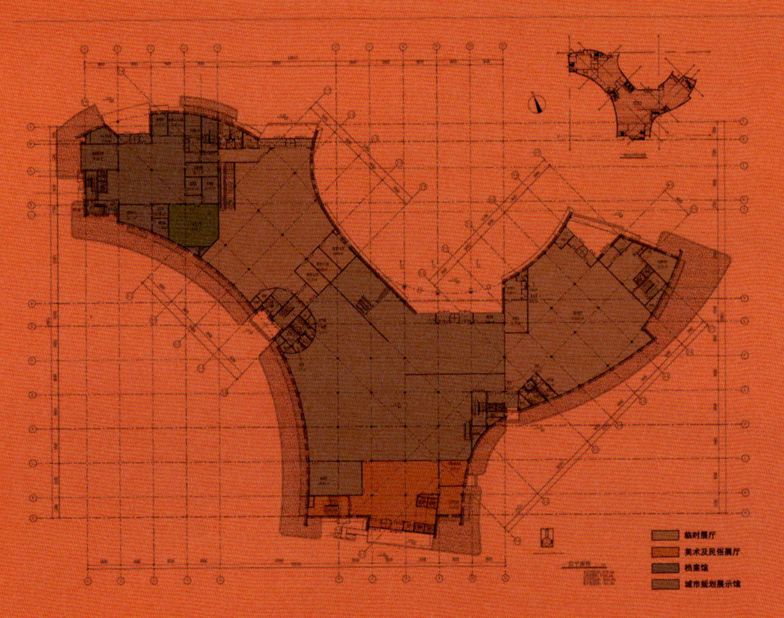

一层平面图 / Ground Floor Plan

旧馆

Old

New

新馆

占地面积： 2560m²
建筑面积： 3745m²
库房面积： 1220m²
层　　数： 主楼 4 层，地下 1 层
设计单位： 新疆维吾尔自治区建筑设计研究院
建成时间： 2014 年 10 月

Site area: 2,560m²
Floor area: 3,745m²
Repository area: 1,220m²
Number of floors: 4 floors and 1 level
Time of completion: October 2014

库尔勒市档案馆有民国时期档案、文书档案、会计档案、声像档案、诉讼档案、婚姻档案等共计18.8万卷（件），现为州级党员干部教育培训（实践）示范基地、库尔勒市爱国主义教育基地、梨城旅游创新发展展示基地。

库尔勒市档案馆新馆坐落于延安路库尔勒展示中心内，与库尔勒规划展示馆、库尔勒民俗文化博物馆合建，档案馆位于西北角，自成一区。馆内布局合理、功能齐全，设有档案查阅中心、现行文件阅览中心、档案特藏室、档案数字化中心、集中归档室、培训室等功能用房，此外还设有“库尔勒地情馆”。

整个建筑外立面为香槟色铝板，与周边建筑与环境协调统一。方案设计取意为“化石”，寓意库尔勒市的过去、现在和未来。

Korla Archives has a total of 188,000 volumes (items) of archives, including archives from the period of Republic of China, as well as administrative, accounting, audio-visual, litigation, and marriage archives. It is now a state-level Party member and cadre education and training (practice) demonstration base, the patriotism education base of Korla City, and the tourism innovation and development demonstration base of the Pear City.

The new building of Korla Archives is located in the Korla Exhibition Center on Yan'an Road. It is jointly built with the Korla Planning and Exhibition Hall and Korla Folk Culture Museum, and seated in the northwest corner of the project as a self-contained area. The Archives has a reasonable layout with complete functions, with functional rooms including the Information Service Reference Center, Prevailing Document Reading Center, Special Archival Collection Room, Archives Digitization Center, Centralized Archiving Room, Training Room, etc. There is also a “Korla Regional Information Hall”.

The facades of the entire building are made of champagne-colored aluminum panels, which are coordinated with the surrounding buildings and the ambient environment. The design concept comes from “fossil”, which implies the past, present and future of Korla.

1 档案馆外景 / The Archives Exterior

2
3 | 4
5

2 展厅 / Exhibition Hall
3 多功能厅 / Multi-purpose Hall
4 档案查阅大厅 / Archives Reference Hall
5 档案整理室 / Archives Filing Room

北屯市档案馆

Beitun Archives

新疆
Xinjiang Production and Construction Corps
生产建设兵团

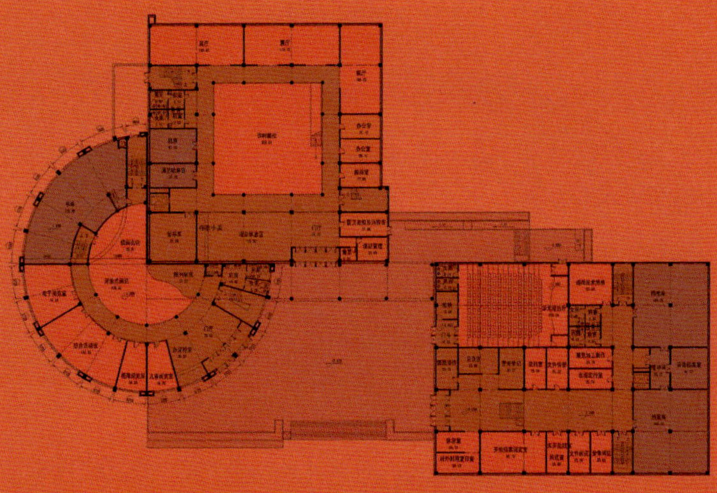
一层平面图 / Ground Floor Plan

占地面积： 8318m²
建筑面积： 4118m²
库房面积： 1685m²
层　　数： 地上 3 层
设计单位： 黑龙江省城市规划设计研究院
建成时间： 2013 年 11 月

Site area: 8,318m²
Floor area: 4,118m²
Repository area: 1,685m²
Number of floors: 3 floors
Time of completion: November 2013

北屯市档案馆位于北屯市南湖新区，十师北屯市行政服务中心西南侧，由北屯垦区综合档案馆与十师188团综合档案馆合建而成，现有馆藏档案7万余卷（件）。

档案馆新馆为新疆北屯博物馆图书馆档案馆工程的重要组成，建筑形体为3个体块，相互组合构成错落有致的现代风格建筑群体组团，成为北屯市城市文化中心。其中档案馆位于东南侧方形体块内，地上3层，每层东侧均为档案库房；西侧部分，一层为服务大厅、展厅、档案查询阅览室、报告厅等对外服务区，二层是办公与数字化用房，整体布局合理，使用方便。

新馆建筑在立面上运用色彩对比和肌理变化体现档案柜的抽象形象，以质朴的线条体现档案管理人的严谨工作作风；在造型设计上，档案馆采用简洁的几何体，简单朴实，实体墙面与通透的玻璃对比鲜明，局部出挑的阳台丰富了立面变化，在质感及色彩上强调现代化氛围中清新、明朗、开放的感觉。

Beitun Archives is located in the Nanhu New District of Beitun City, on the southwest side of the Administrative Service Center of the 10th Division Beitun City. It is a joint project of Beitun Reclamation Area Comprehensive Archives and the Corp 188 Comprehensive Archives of the 10th Division, and now has a collection of more than 70,000 files/items of archives.

The new building of the Archives is an important component which also comprises the Museum and Library of Beitun, Xinjiang. The building is composed of three masses, which are combined to form an irregularly arranged modern-style architectural cluster, and become the urban cultural center of Beitun City. The Archives is located in the mass in the southeast side, with 3 floors aboveground. The east part of each floor is for archives repositories; the west part of 1F accommodates the Service Hall, Exhibition Hall, Archives Reference and Reading Room, Lecture Hall and other public service areas; and the west part of 2F is for Offices and Digital Rooms. The overall layout is reasonable for easy utilization.

On the building's facades, color contrast and texture changes are used to reflect the abstract image of archives cabinets, and simple lines used to represent the rigorous work style of the archival managers. The design employs simple and natural geometries; distinct contrast between the solid wall surfaces and transparent glass, as well as the partially overhanging balconies enrich the facade with changes, emphasizing the feeling of freshness, clarity and openness with textures and colors in a modern atmosphere.

1 档案馆外景 / The Archives Exterior

十师北屯市档案馆

2
3
4

2 入口大厅 / Entrance Lobby
3 档案查阅室 / Archives Reference Room
4 展厅 / Exhibition Hall

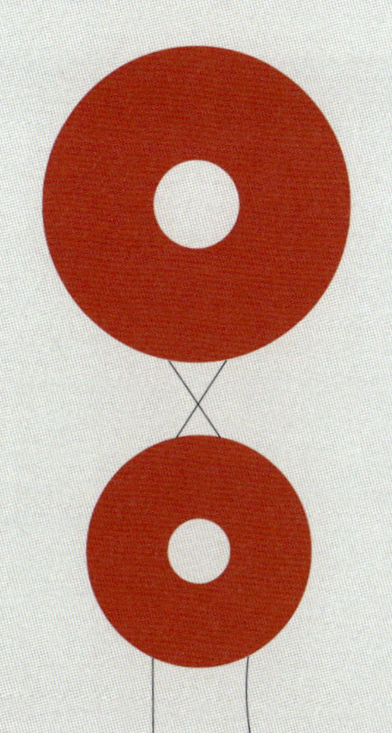

中西部地区县级综合档案馆建设代表实例

Demonstration Cases of Construction for the County-level General Archives in Central and Western China

河北省 | Hebei Province

迁安市档案馆

Qian'an Archives

山西省 | Shanxi Province

长治县档案馆

Changzhi Archives

永济市档案馆

Yongji Archives

内蒙古自治区 | Inner Mongolia Autonomous Region

阿拉善左旗档案馆

Alxa Left Banner Archives

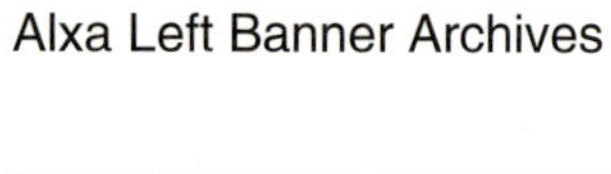

赤峰市档案馆

Chifeng Archives

托克托县档案馆

Togtoh Archives

扎兰屯市档案馆

Zhalantun Archives

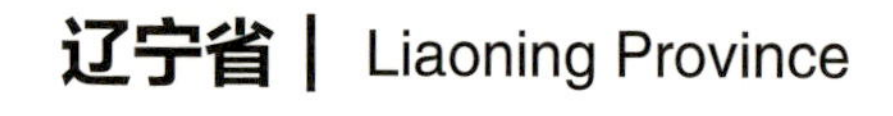

辽宁省 | Liaoning Province

康平县档案馆

Kangping Archives

西丰县档案馆

Xifeng Archives

吉林省 | Jilin Province

德惠市档案馆
Dehui Archives

黑龙江省 | Heilongjiang Province

漠河县档案馆
Mohe Archives

通河县档案馆
Tonghe Archives

安徽省 | Anhui Province

当涂县档案馆
Dangtu Archives

金寨县档案馆
Jinzhai Archives

石台县档案馆
Shitai Archives

福建省 | Fujian Province

光泽县档案馆
Guangze Archives

建宁县档案馆
Jianning Archives

南靖县档案馆
Nanjing Archives

江西省 | Jiangxi Province

诏安县档案馆

Zhaoan Archives

奉新县档案馆

Fengxin Archives

上饶市档案馆

Shangrao Archives

河南省 | Henan Province

余干县档案馆

Yugan Archives

郏县档案馆

Jiaxian Archives

清丰县档案馆

Qingfeng Archives

湖北省 | Hubei Province

宜阳县档案馆

Yiyang Archives

当阳市档案馆

Dangyang Archives

郧西县档案馆

Yunxi Archives

湖南省 | Hunan Province

长阳土家族自治县档案馆
Archives of Changyang Tujia Autonomous County

洪江市档案馆
Hongjiang Archives

祁阳县档案馆
Qiyang Archives

广西壮族自治区 | Guangxi Zhuang Autonomous Region

双峰县档案馆
Shuangfeng Archives

北海市海城区档案馆
Haicheng District Archives,Beihai

防城港市港口区档案馆
Gangkou District Archives, Fangchenggang

贵港市港北区档案馆
Gangbei District Archives, Guigang

贵港市港南区档案馆
Gangnan District Archives, Guigang

合山市档案馆
Heshan Archives

重庆市 | Chongqing Municipality

南宁市良庆区档案馆
Liangqing Distict Archives, Nanning

容县档案馆
Rongxian Archives

潼南区档案馆
Tongnan District Archives

万盛区档案馆
Wansheng District Archives

武隆区档案馆
Wulong District Archives

秀山县档案馆
Xiushan Archives

四川省 | Sichuan Province

永川区档案馆
Yongchuan District Archives

酉阳县档案馆
Youyang Archives

洪雅县档案馆
Hongya Archives

贵州省 | Guizhou Province

邻水县档案馆
Linshui Archives

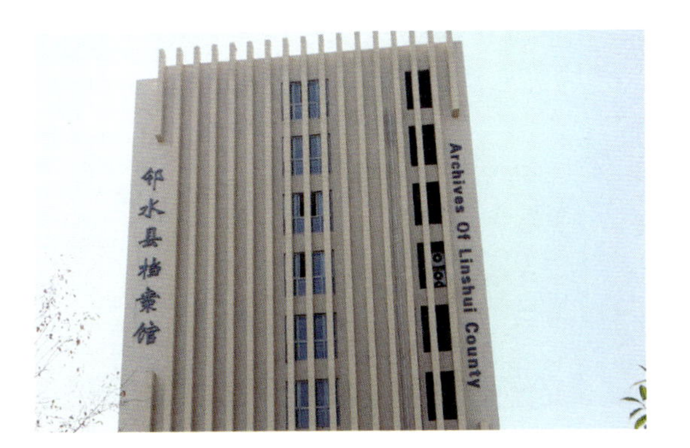

炉霍县档案馆
Luhuo Archives

施秉县档案馆
Shibing Archives

云南省 | Yunnan Province

沿河县档案馆
Yanhe Archives

德钦县档案馆
Deqin Archives

泸西县档案馆
Luxi Archives

南涧县档案馆
Nanjian Archives

师宗县档案馆
Shizong Archives

香格里拉市档案馆
Shangri–La Archives

西藏自治区 | Tibet Autonomous Region

林周县档案馆
Linzhou Archives

米林县档案馆
Milin Archives

陕西省 | Shaanxi Province

韩城市档案馆
Hancheng Archives

麟游县档案馆
Linyou Archives

略阳县档案馆
Lueyang Archives

商洛市商州区档案馆
Shangzhou District Archives, Shangluo

甘肃省 | Gansu Province

宜君县档案馆
Yijun Archives

靖远县档案馆
Jingyuan Archives

西和县档案馆
Xihe Archives

青海省 | Qinghai Province

海晏县档案馆
Haiyan Archives

河南县档案馆
Henan Archives

宁夏回族自治区 | Ningxia Hui Autonomous Region

灵武市档案馆
Lingwu Archives

新疆维吾尔自治区 | Xinjiang Uygur Autonomous Region

阿克苏市档案馆
Aksu Archives

库车县档案馆
Kuqa Archives

新疆生产建设兵团 | Xinjiang Production and Construction Corps

142 团档案馆
Corp 142 Archives

乌鲁木齐垦区档案馆
Urumqi Reclamation Area Archives

后记

随着改革开放的不断深入，我国档案事业蓬勃发展，尤其是大量档案馆新馆建成并投入使用，让档案馆面貌焕然一新。特别是近10年来，在中央财政的有力支持下，中西部地区县级综合档案馆馆舍建设在原先的基础上有了质的飞跃，普遍达到了《档案馆建设标准》和《档案馆建筑设计规范》的基本要求，能够满足履行《中华人民共和国档案法》规定的法定职责的基本需要。中西部地区县级综合档案馆建设的加快，实现了我国档案事业的区域协调发展，也使越来越多的群众能够就近享受到最基本、最便利的档案公共服务。

建兰台，筑丰碑，织蓝图，歌盛世。在新时代，各级综合档案馆要坚持以习近平新时代中国特色社会主义思想为指导，主动适应国家治理体系和治理能力现代化的深刻变革，立足档案馆的资源优势和专业优势，紧紧围绕爱国主义教育基地、档案安全保管基地、档案利用服务中心、政府公开信息查阅中心和电子档案备份管理中心等5个方面的基本功能，深化改革、开拓创新，大力提升档案馆业务建设规范化、科学化水平，更好发挥档案工作的基础性、支撑性作用，在服务党和国家中心工作、服务经济社会发展、服务人民群众中努力谱写档案事业发展新篇章。

本书在编撰过程中，得到了全国各级档案局、馆的大力支持以及有关同仁的密切配合，本书选用图片及档案馆文字简介初稿均由各地档案馆提供，张倩（河北）、赵正之（山西）、张晔（内蒙古）、刘薇薇（辽宁）、姜福涛（吉林）、祝博强（黑龙江）、范武（安徽）、陈亮（福建）、叶修斌（江西）、范璟玉（河南）、王尹芹（湖北）、熊燕（湖南）、高力（广西）、叶志勇（海南）、李海燕（重庆）、赵怡（四川）、胡朝强（贵州）、陈建东（云南）、达瓦次仁（西藏）、赵辉（陕西）、杨生吉（甘肃）、康鹏君（青海）、温鹏（宁夏）、王国华（新疆）、张玮（兵团）等地方档案局、馆的同志在组稿过程中做了很多工作，黄习习、王鹏、徐慧敏3位北京建筑大学的研究生对档案馆文字简介部分做了专业的修改、加工和润色，在此谨向所有给予此书支持和帮助的同志表示由衷的感谢！

受版面所限，本书仅收录了140家档案馆，相对于列入《中西部地区县级综合档案馆建设规划》的2000余家档案馆只是很少的一部分，对提供了材料却因各种原因未能收录的档案馆，我们只能表示歉意。由于时间仓促、水平有限，本书难免存在不足之处，敬请各位读者不吝指正。

编　者

2019年9月

Afterword

The archives cause of China has been growing vigorously along with the constant advancement of reform and opening-up. Particularly, a large number of new archives buildings already completed and put into operation give archival facilities a new look. In the past 10 years in particular, under the full support from the central government, the county-level general archives in central and western China have witnessed a qualitative leap, as the majority of them reach the fundamental requirements as set forth in *Construction Standards for Archives and Design Code for Archives Buildings*, and fulfill the statutory duties as stipulated in the Archives Law of the People's Republic of China. The acceleration in the construction of county-level general archives in central and western China has facilitated the regional coordinated development of archives cause in China, so that more and more people have access to essential and convenient public archival services.

Responding to the call of building classics repository, erecting monuments, drawing up blueprints, and singing in praises of the flourishing age, in the new era, general archives at all levels must do the following work. That is to adhere to "Xi Jinping Thought on Socialism with Chinese Characteristics for a New Era", and take the initiative to adapt to the profound changes in the national governance system and the modernization of governance capabilities; capitalize on the resource advantages and professional advantages of the archival facilities, and closely stick to the five basic functions, including a base for patriotism education, a base for archives custody, a center of archives information and resources utilization, a center of government public information consultation, and a center of electronic file backup management; deepen the reform, keep exploring and innovating, and put great efforts to make the business construction of archives more standard and scientific; give full play to the fundamental and supportive role of archival work; and strive to make breakthroughs of the archives undertaking while serving the Party and the main tasks of the state, the public, Economic and Social development.

Special thanks should go to the archives administrations and archival facilities at all levels throughout the country for their generous support during the compilation of this book, to the colleagues for their close cooperation, to the archival facilities for their provisions of pictures and introduction text draft in the book. The thanks also go to Zhang Qian (Hebei), Zhao Zhengzhi (Shanxi), Zhangye (Inner Mongolia), Liu Weiwei (Liaoning), Jiang Futao (Jilin), Zhu Boqiang (Heilongjiang), Fan Wu (Anhui), Chen Liang (Fujian), Ye Xiubin (Jiangxi), Fan Jingyu (Henan), Wang Yinqin (Hubei), Xiong Yan (Hunan), Gao Li (Guangxi), Ye Zhiyong (Hainan), Li Haiyan (Chongqing), Zhao Yi (Sichuan), Hu Chaoqiang (Guizhou), Chen Jiandong (Yunnan), Dawa Tsering (Tibet), Zhao Hui (Shaanxi), Yang Shengji (Gansu), Kang Pengjun (Qinghai), Wen Peng (Ningxia), Wang Guohua (Xinjiang), Zhang Wei (Corps) and other peers from local archives administrations and archival facilities, who have done a lot of work during contribution soliciting, as well as three graduate students from Beijing University of Civil Engineering and Architecture Huang Xixi, Wang Peng and Xu Huimin, who have revised, edited and polished the introduction to archival facilities as professionals. Thank you all again for your kind support and assistance made to the book!

Due to limitation of pages, this book presents only 140 archival facilities, a very small portion of total 2,000 plus listed in the *Construction Planning for County-level General Archives in Central and Western China*. It is a pity that some archival facilities, though with relevant information submitted, could not be included into this book for certain reasons and I feel sorry for that. The limit of my time and my knowledge makes it hard to be one hundred percent correct. Your suggestions will be appreciated.

Editor

September 2019